W9-DGE-708

NONTECHNICAL GUIDE TO

ENERGY RESOURCES

◆

AVAILABILITY, USE AND IMPACT

◆

NONTECHNICAL GUIDE TO

ENERGY RESOURCES

AVAILABILITY, USE AND IMPACT

BEN W. EBENHACK

PennWell Books

PENNWELL PUBLISHING CO.
TULSA, OKLAHOMA

PennWell Publishing Company

1421 South Sheridan/P.O. Box 1260

Tulsa, Oklahoma 74101

Library of Congress Cataloging-in Publication Data

Ebenhack, Ben W.

Nontechnical guide to energy resources: availability, use and impact/

Ben W. Ebenhack.

p. cm.

Includes bibliographical references and index.

ISBN 0-87814-434-X

1. Power resources.

2. Energy development — Environmental aspects.

I. Title.

TJ163.2.E23 1995

333.79—dc20

Printed in the United States of America

1 2 3 4 5 98 97 96 95 94

I would like to
dedicate this work to
MJ, BJ, and
my entire family.

Contents

ACKNOWLEDGEMENTS xv
INTRODUCTION xvii
An Historical Context xx
Costs of Energy xxiii
Availability xxiii
Rational Choices xxvi
Energy Consumption in the Nonindustrial World xxvii
Tracing the Path of Energy xxix

CHAPTER 1

ENERGY SOURCES, RESOURCES, AND RESERVES 1
Definitions 1
Resources and Reserves 3
Renewable Resources 7
Fossil Fuels 9
Nuclear Energy 10
Abundance and Availability 10
Combustion Fuels 12
Coal 12
Occurrence 13
Rank and Quality 14
Oil and Gas 16
Occurrence 17
Oil Markets 19
Future Production Potential 26
Unconventional Oil and Gas Resources 27
Kerogen 28
Biomass 28
Noncombustion Sources 30
Solar Power 30

Wind Power .32
Hydropower .33
Geothermal Power .33
Nuclear Power .33
Conservation .33
Conclusions .34

CHAPTER 2

ACQUISITION .39
Combustion Fuels .39
Coal .39
Exploration .39
Extraction .40
Underground Mining Techniques .42
Strip Mining .45
Coalbed Methane .46
Oil and Gas .46
Exploration .46
Meaning of a Commercially Successful Discovery Well .49
Drilling .50
Drilling Rig .52
Drilling Procedure .54
Drilling Problems .56
Post-Drilling Evaluations .59
Porosity .62
Saturations .63
Completions .64
Stimulation .67
Production .68
Secondary and Enhanced Oil Recovery .70
Onshore and Offshore .74
Oil Shale or Kerogen .76
Exploration .76
Extraction .77

Biomass 79
Exploration and Fuelwood Deficits 79
Biomass Agriculture 82
Biogas 85
Energy from Waste 86
Noncombustion Sources 87
Solar Power 87
Exploration 87
Active Solar 89
Passive Solar 91
Wind Power 93
Hydropower 93
Exploration 93
Acquisition 95
Wave and Tidal Energy 96
Geothermal Power 97
Exploration 97
Production 98
Open and Closed Loop Production Systems 98
Types of Geothermal Reservoirs 99
Ocean Thermal Electric Conversion 100
Nuclear Power 100
Conservation 104
Exploration 104
Acquisition (Implementation) 105

CHAPTER 3
TRANSPORTATION AND STORAGE 109
Combustion Fuels 110
Coal 110
Pipelining 113
Oil and Natural Gas 114
Tankers 118
Natural Gas Transportation 121

Biomass 123
Noncombustion Fuels 125
Solar Power 125
Storage 127
Transport 130
Wind Power 131
Hydropower 131
Geothermal Power 133
NUCLEAR POWER 134
Transmission of Electricity 135
Storage 137
Electrochemical storage 137
Kinetic Energy Storage 140
Compressed Air 141

CHAPTER 4

CONVERSION AND END-USE 143
Combustion Fuels 143
Coal 143
Coal Combustion 146
Coke Production 147
Gasification 147
Liquefaction 148
Oil and Gas 150
Natural Gas Processing 150
Crude Oil Refining 151
The Refinery 153
Fuel Requirements 158
Kerogen 161
Biomass 161
Conversion to Secondary Fuels 165
Noncombustion Sources 169
Solar Power 169
Thermal Applications 169

Photovoltaic Applications .170
Solar Thermal Electrical Conversion .170
Wind Power .171
Slow and Steady .172
Wind for Transport .173
Hydropower .174
Geothermal Power .175
Nuclear Energy .175
Direct Fission Reactions .175
Enhanced Fission Reactions .177
Controlled Fusion .177
Electricity .179
Generating Electricity .179
Use of Electricity .180
Conservation .181
Abstinence .181
Efficiency Improvements .182

CHAPTER 5

THE IMPACT OF ENERGY USE .187
Combustion Fuels .188
Greenhouse Warming .188
Acid Rain .190
Coal .191
Acquisition .191
Mine Safety .193
Transport .196
Air Pollution .197
Acid Rain and High Sulfur Coals .198
Carbon Dioxide Emissions .200
Oil and Gas .201
Acquisition .201
Leaks and Spills .202
Shipping Spills .206

Vehicular Fuel Demand .210
Kerogen .211
Biomass .212
Biomass in Nonindustrialized Countries .215
Methane from Biomass .216
Ethanol Production .217
Energy from Waste .218
Noncombustion Sources .218
Solar Power .218
Passive Solar .219
Solar Thermal Electric Conversion and Indirect Heating .220
Photovoltaic .220
Wind Power .221
Hydropower .223
Disruption of Local Ecology .223
Changing Patterns .224
Tidal and Wave Power .225
Geothermal Power .225
Ocean Thermal Electric Conversion .226
Nuclear Power .227
Myths .227
Hazards of Nuclear Power .230

CHAPTER 6

ENERGY CHOICES .235
Abundance and Availability .236
Supply and Demand Issues .236
Resources and Renewability .238
Sustainable Firewood Production .239
The Future of Oil and Gas Production .242
Coal: The Backstop Energy Resource .243
The Nondepletable Sources .243
Nuclear Power .244
Acquisition .245

Coal .245
Oil and Gas .246
Kerogen .248
Biomass .249
Solar .250
Wind .251
Hydropower .251
Geothermal .252
Nuclear Power .252
Transportation and Storage .253
Combustion Fuels .254
Electricity .254
Conversion and End-Use .255
Combustion .256
The Role of Conservation .257
Voluntary Energy Thrifty .257
Building Standards .258
Appliances .259
Savings from Decreased Raw Material Consumption .260
Conservation Policy Options .260
Balancing the Impacts .261
The Value of Externalities .263
The Environment of Renewability .266
The Closed Carbon Cycle .268
Economic Restraints .269
The Cost of Fuel Efficiency Standards .269
The Environment and Aesthetics .270
Economies of Scale .272
Conclusion .274
Cost and Benefit Comparison .278

APPENDIX .281

INDEX .287

ACKNOWLEDGEMENTS

I appreciate the input of numerous colleagues and an even greater number of students, past and present. In particular, I wish to thank my colleagues: Karen E. Fields, Joseph Inikori, C. Anthony Bush, Dave Chappell, C. Augustus Nwaubani, Wilson Ogbomo, Morris Pierce, Dave Weimer, and Harvey Palmer. Students particularly helpful were: Shawn King, Gillian Philips, Britt Bollinger, and Densie Brien.

INTRODUCTION

Over the past two decades, we have begun to think of energy as a problem, and we tend to see that problem in various, sometimes conflicting, ways: There is too little of it; increasing energy production injures our environment; the by-products of its use add insult to the injury; we ought to conserve; we require more energy to provide the benefits of modern society; energy prices are too low, or too high; and so on. In this manner, we have zigzagged our way through a tangle of hot debates over one aspect or another of what we think of as *the energy problem.* Some of those strands lead back to production, and some to use, but all concede the natural importance of energy in our lives. Hence the need to think about energy in a comprehensive way.

This book is intended not only to provide facts about energy resources but a rational analyses of the impacts caused by their use. It is meant to explore as well the role that energy plays in our lives and beings—from the family meals cooked over gas flames in many urban kitchens of North America and over wood flames in those of Africa, to the stories narrated worldwide by means of electronic images scrambled, sent through space, and reassembled before our eyes.

More fundamentally, energy is essential to life itself—from the minute assimilation of light energy by phytoplankton to the massive metabolic consumption required to maintain body temperatures and neural and physical activity among large, warm-blooded creatures. Even the internal energy demands of the largest and most active animals are dwarfed in comparison to the external energy demands of a typical individual in a modern industrial society. The inventor Buckminster Fuller referred to energy as modern society's "slave," providing the comforts, services, and entertainment which could only have been afforded by vast numbers of people and animals in pre-industrial societies.[1]

Humanity's use of external energy sources propelled our way to the top of the food chain. Our ancestors employed energy to harden spear points and to broaden their diets by cooking food. It seems we seek to secure our niche in the evolutionary framework by striving to employ ever greater power, even in the pursuit of trivial activities.

Commercial and large-scale energy acquisition efforts have long been enterprises attracting pioneers who have employed exotic and sometimes daring means to recover more of the energy resource. From long before the industrial revolution, when the Chinese *kicked wells down* several hundred feet in search of water, through the early Welsh coal mining, to modern drilling of oil wells nearly six miles deep, mining and drilling have tapped into the immensity of the earth's forces as well as its resources. How do geoscientists locate underground geothermal and uranium resources, as well as fossil fuels?

To this day, many people are surprised to learn that oil and gas occupy the tiny pores between grains in sandstones (or other porous rocks) rather than in huge, subterranean caverns. In that case, how is it extracted? In fact, petroleum producers are generally fortunate to extract one-third of the original volume of oil found in a reservoir. To enhance the rate of extraction and the ultimate recovery, some extraordinary techniques have been employed. Along with more conventional technologies, this book will lead the way through explosive fracturing—from early drillers *shooting wells* with nitro glycerine to the DOE's project Gas Buggy (detonating atomic bombs in wellbores!). This book also describes a gusher and why its modern-day equivalent, a blowout, is not the successful culmination of drilling. (The 1960s movie *Hellfighters,* starring John Wayne, offered a surprisingly accurate depiction of blowouts, and the image was brought sharply home to the general public by the aftermath of the Persian Gulf War.)

Answers are provided to the most basic questions: What holds up the roof of an underground coal mine? Why did miners carry canaries into coal mines? What is the lever arm device teeter-tottering in oil fields one passes on the highway? More practical questions will be answered: What determines whether to strip-mine or mine underground? How much energy does wind provide? What energy resources can be more extensively tapped under existing economic conditions, in different countries? Information and suggestions for approaching the questions that have no answers will also be presented: What energy alternatives offer the most hope for the coming century? Is coal an environmentally better option than nuclear power? When will oil and gas resources be economically depleted?

Many of these questions do not have simple answers. There is an immense array of energy uses. Some require intense energy for short periods of time; some are slow and steady tasks. Some are discretionary, while others are essential to life. Similarly, each of the different energy resources on which humans can draw have different properties. Some energy resources will be practically exhausted someday. Some forms of energy require large investments. Some energy resources are only available in specific locations or at specific times. Furthermore, some energy forms can be very benign in their final use, but how were they produced and delivered to the consumer? Many of the problems associated with energy are invisible to the user. Since there is a broad range of needs and resource characteristics, the best answer to the energy-use question is probably a mixture. A robust form of analysis is necessary to encompass the entire range of advantages and disadvantages accruing to all of the energy options throughout their production cycles.

In general, *Nontechnical Guide to Energy Resources: Availability, Use and Impact* offers a factual understanding of what energy resources are, how they are produced and utilized, and what their known environmental impacts are. The book also raises some questions for consideration and offers some new perspectives. Whether you're interested in an energy-producing career, assessing environmental impacts, or planning policies to assure adequate and reliable energy resources for the future, this text offers a starting point.

AN HISTORICAL CONTEXT

Energy has played a pivotal role in the evolution of society. Humanity's earliest energy source was animal power—the human animal first. However, the domestication of animals to serve as external energy sources probably came much later in prehistory than the domestication of fire. Indeed, the dominant energy source for Americans was animal power even until the middle of the 19th century. Some authors include food as an energy source. Food provides the internal energy for the survival of all creatures. The ability, though, to harness external energy sources helped to set humans apart from other animals and to ensure the survival of the species.

The first external energy source was firewood. Recognizing the importance of harnessing energy, the early Greeks developed a creation myth about the acquisition of fire. According to the myth, the god Prometheus, charged with doling out special gifts and abilities to each species, realized when he came to humans that the gifts of speed, strength, flight, etc., were already given to others. In desperation, he gave humans fire. This angered the other gods who, taking a very proprietary attitude towards fire, condemned our benefactor to eternal torture. Literature aside, though, it is interesting to note that the Greeks depicted an external energy source, rather than intelligence or an opposable thumb, as the gift which sets humanity apart from other animals. Indeed, fire has provided for human survival and development ever since.

Firewood and domestic animal power were the dominant energy sources for humanity for hundreds of millennia. Only as wood became scarce did a search for new energy alternatives begin. The loss of forests in northern Europe is generally believed to have driven a shift to coal in the few centuries immediately preceding the Industrial Revolution.[2] Coal was able to provide immensely more energy than firewood, literally fueling industrialization.

In the United States, during its early years, whale oil met the need for a household fuel to cook meals and light the evenings. Another shortage, this time of oil from dwindling whale populations, inspired the quest for a new energy source. Investors hired an itinerant railroad worker to drill a hole into the earth in an effort to find commercial quantities of petroleum which seeped to the surface in some places.

The oil boom quickly displaced other energy supplies. Petroleum was a more versatile, transportable, and efficient fuel than had been known before in quantity. Unlike sources that required harvesting or mining, oil flowed from wells in the ground. The companies selling the product eagerly sought new markets for their burgeoning production, and consumers were quick to respond. Initially, kerosene was the petroleum product of greatest value which was used to meet cooking and lighting needs. In the beginning of the 20th century, though, the introduction of the automobile provided a vast new market for the liquid fuel. Gasoline and diesel production soon accounted for more than half of the petroleum used in the United States.

In the early days of the petroleum industry, natural gas was generally a waste product and was burned away on site. As pipelines spread across America, it became feasible to deliver gas to consumers. Natural gas carved a large market niche for cooking, heating, and lighting which had been oil's original *raison d'être.* Some experts believe that the market share of clean-burning, efficient gas will continue to grow and surpass other sources in the coming decades.[3]

By 1990, energy was impinging on the world's consciousness for another reason: rising environmental awareness. Attention focused heavily on carbon dioxide emissions and acid rain which can result from the use of fossil fuels. But what are the alternatives? Traditional biomass combustion (i.e., firewood or charcoal) actually has all of the disadvantages of most fossil fuels, and newer biomass technologies have their own problems, including present day emissions of carbon dioxide and other pollutants.

Hydropower generation has been around for centuries, in one form or another, yet much of the high quality hydroelectric potential not presently harnessed is extremely remote (such as in northern Alaska). Solar energy is abundant, but intermittency and low conversion efficiencies prevent it from being economically competitive, except for remote or minute-scale applications. Wind power, also utilized for centuries, is very intermittent, and population centers have not grown up around the most ideal sites for this energy resource—where there is a constant, howling wind. Geothermal energy can be very clean, and has a virtually inexhaustible resource base, but occurrences of commercially viable potential are relatively rare. Nuclear fission produces neither greenhouse gasses nor acid rain but leaves waste products which remain dangerously radioactive for centuries to come. Controlled nuclear fusion is at best a resource for the future, at worst, a pipe dream.

A final energy source is conservation. Conservation can come in the form of abstinence from discretionary energy use or employing more efficient technologies when energy is used. In the latter sense especially, it permits deriving the same benefits with less energy consumed. The energy conserved is then made available for other tasks. It is hard to think of anything intelligent to say against conservation, except that it seems easier to advocate than to practice. Wise people have urged frugality throughout the ages, and this author would advocate serious conservation efforts in the industrialized nations (especially the United States and Canada). Nevertheless, conservation is not a total solution. Even if a dramatic 50% reduction in energy consumption were achieved in America, demand would continue to be approximately five times the total energy production provided by solar, geothermal, wind, and hydropower combined. Furthermore, burgeoning populations and the current desperately low energy use among perhaps two billion people in lower-income countries dictate that, at best, the world can slow the growth in energy consumption to about 2% per year in the coming decades.[4]

COSTS OF ENERGY

Carbon dioxide buildup, the *greenhouse effect*, acid rain, and the specter of disaster raised by nuclear power accidents all suggest the possibility that our *energy slaves* may rise up against us. The industrialized world has become dangerously dependent on massive energy consumption. This dependency currently makes the economies of most industrialized societies vulnerable to the disruption of essential imports, as well as incurring serious environmental costs. Many of the environmental costs associated with various energy sources are well publicized. Yet the situation is not as simple as eliminating villainous fossil fuels or nuclear power. There are problems associated with every energy form. The combustion products of all the fuels we burn create problems and have costs in terms of human life and environmental damage in their acquisition and delivery as well. Firewood seems benign to many but still emits carbon dioxide and pollutants; and in some parts of the world, the use of firewood contributes to deforestation and respiratory illness. Solar and wind power facilities can generally not compete economically with the combustion fuels and have a range of impacts of their own.

AVAILABILITY

Beyond the costs of environmental damage are the problems associated with availability. All commercially usable fossil fuels (especially the cleaner oil and gas) will no doubt be effectively exhausted some day, unless a cheaper and more efficient energy source is discovered to replace them. Solar and wind resources are intermittent and demand sophisticated storage technologies if they are to become stand-alone energy providers. Economics plays a strong role in availability. Consumers generally opt for cost advantages over environmental or long-term production advantages. Neo-classical economics would refer to this tendency as *rational choices.* However, consumer preferences which may not be rigorously rational can determine the demand also, and thus the market for energy. While it is easy to condemn fossil fuels and to dream of solar energy (much as the author's generation dreamed of utopian societies fueled by clean and abundant nuclear energy), the responsibility for the failure to develop clean and inexhaustible energy sources must be shared by the consumer. People can blame governments for not allocating enough money for research on solar, geothermal, and wind power. People can blame companies for not offering clean energy resources and consumption appliances. But, through the energy-rich 1980s, cars which achieved twice the average fleet fuel economy sold so poorly that automotive manufacturers became convinced that they could not sell cars which would meet the government's (modest) goals for improved fuel efficiency. Consider then how many Americans would buy a solar-powered car, with a 30 mile per hour cruising speed and a range of 100 miles.

Energy supplies tend to appear as a given in modern society, taken for granted until some disruption brings them into focus. In 1973, Americans suddenly became aware of energy supplies when Arab nations enacted an embargo on petroleum shipments to the United States (and the Netherlands) in protest of American support for Israel. The decade-old Organization of Petroleum Exporting Countries (OPEC) flexed its muscle, imposing price hikes in the wake of the embargo. Prices which had been static for 20 years rebounded dramatically. The first serious energy conservation efforts since World War II in the largest consuming country was spurred by fears of shortage and "running out of oil and gas." Shocked, Americans talked of sinister plots, directing suspicion at Arabs and oil companies alike—suspicions that oil tankers were anchored outside of ports, withholding their cargo to drive yet another price increase.

The rest of the world saw a somewhat less startling picture, though often no less disruptive. In western Europe, gasoline prices had been several times higher than U.S. prices for years. Heavy taxes had long been levied on gasoline, thus vehicular demand never reached American levels. Still, Europeans and Japanese, even more dependent on imported energy than Americans, had cause for anxiety. "Energy security" became a battle cry in government planning around the world.

Oil companies ran budget-projection case studies reflecting the price of oil increasing from the 1979 prices of $36/barrel to $60/barrel by 1990. Salesmen maintained that solar water heaters would pay for themselves in savings, based on energy prices continuing to escalate at the rate seen in the 1970s. Many energy alternatives teetered on the brink of commercial viability. Coal, by far the most abundant of the fossil fuels, could be converted into a more broadly marketable (and potentially environmentally superior) gaseous or liquid fuel. Oil shale, also tremendously abundant, was another resource which could be converted into liquid *synthetic crude*. Dozens of nuclear power plants were commissioned for construction.

Millions of Americans were shocked to hear that oil and gas reserves could only meet demand for a few decades. People were understandably skeptical. How could America run out of oil and gas so soon? Had there been some conspiracy to hide dwindling reserves, or was the current shortage a fraud? In fact, America would not run out of oil and gas in the foreseeable future, nor was there any great conspiracy. To some degree, the *shortage* was a problem of terminology. Expressing a reserve by the number of years it would meet demand is really mixing apples and oranges. It ignores the realities of how reserves are defined and evaluated. It can be a useful relativistic way to contrast the finite resource base of a fossil fuel with the billion-year resource base of solar energy. Even in such a comparison, though, care must be taken not to mix resource base with reserves, or to mix a resource's present rate of consumption with total demand. It is essential, then, to understand the true meaning of all of these terms and more, as well as to understand the fundamental technologies of energy development and production.

In 1981 the energy shortage of the 1970s seemed suddenly

to reverse itself, and so did the consumers' response. The world saw an oil glut and tumbling energy prices on the heels of a seemingly endless shortage. Nuclear power plant construction screeched to a halt. Conservation lost popularity. While small cars sold in the late 1970s based on their *EPA mpg* rating, by the end of the next decade the major selling points were horsepower, size, four-wheel-drive, and acceleration from 0 to 60 miles per hour. Commitment to conservation has seemingly been very responsive to the prices faced by the consumers.

The *oil crisis* forced the public to recognize that consumer demand depletes oil and gas reserves, and that those resources will inevitably be exhausted in the future. It popularized contrasting *renewable* energy resources with *non-renewable* resources. Some resources can be depleted by human actions (such as firewood) but have much higher rates of renewal than the fossil fuels. Other resources (such as solar and wind power) do not have reserve accumulations from which to draw, but they cannot be depleted by being harnessed for human use. The term *renewable* came into use to address the concern of running out of energy. Somehow, renewability came to be considered as synonymous with *environmentally benign* in popular perception. The connection is referred to regularly enough to merit considering its validity.

RATIONAL CHOICES

Economists use the expression *rational choices* to describe decisions made by perfectly informed consumers in order to maximize their own benefit. Misinformation and lack of information are two prevalent reasons why consumers depart from rational choices. Most consumers tend only to see the end use of energy—not its acquisition, transportation, or conversion. The use of massive solid fuels, such as coal and biomass, consumes large amounts of energy and tends to have a high incidence of accidents in the acquisition and transportation phases. Indeed, there are problems with all choices. Although diesel fuels emit visible and odoriferous products upon combustion, they incur less environmental costs than gasoline in the refining process. Hydroelectric power generation, which is noteworthy for its cleanliness and utter lack of airborne pollutants, still raises environmental concern for its potentially radical disruption of sedimentation patterns and local ecology. No one seems to want to bear the risks of having a nuclear power plant or waste site near his or her home, but the airborne pollutants of coal-fired power plants spread their hazards to the whole world. There is much talk of the long half-lives of radioactive wastes, yet combustion products polluting the atmosphere have even longer lives. Consider too the environmental favorite, solar energy; photovoltaic cells and the storage batteries in common use contain toxic compounds, creating a disposal problem at the ends of their lives. While all energy sources have social and environmental costs, no one in the industrialized world seems willing to curtail discretionary energy use or to pay high prices for their energy.

Another assault on rational choices arises from confusing issues. Aesthetic issues are easily confused with environmental ones. Hated offshore drilling is responsible for much less oil spillage than tankering imports, yet no one seems willing to have an offshore platform or rig within sight of his or her expensive ocean-front property. Therefore, the environmental impact of drilling off the coast of the United States may be positive, since imported oil is the present, cost-competitive alternative, which is increasing due to limits on domestic oil production.

ENERGY CONSUMPTION IN THE NONINDUSTRIALIZED WORLD

What is the primary energy source for the most people in the world? Biomass (firewood and charcoal). Biomass is certainly not the largest provider of energy, but it provides energy for more people than any other resource.[5] Biomass has a nice sound to it, and many would advocate its use unconditionally, because of the *closed carbon cycle* and renewability. But if one considers sub-Saharan Africa, where biomass is the chief energy resource and where fuelwood demands account for as much as 80% of deforestation, it is clear that biomass neither exhibits a closed carbon cycle nor is adequately renewable. In fact, both of these attributes are not absolute but are best described as rate functions, as will be seen in the final chapter.

Energy is even more a burning issue in the Third World than in the industrialized world because of the lack of adequate supplies—and options. Economic constraints cause energy resources produced in nonindustrialized countries to be destined for export to the markets of the industrialized world, while local residents remain dependent on firewood and charcoal.

This book will explore the technical and economic bases of the Third World energy paradox, along with various proposed solutions and their potential. Interestingly, though, energy's role in development and relief is often hidden. In the nonindustrialized countries, the per capita energy consumption can be as low as 1.2% of the per capita consumption in the United States.[6] This means that when we add up such commonplace uses as heated and cooled interiors; well-lit homes, businesses, streets, and ballgames; sterilized medical implements; and frozen food; 1 American consumes as much energy as 80 Mozambicans. Most development experts agree that the energy consumption of Americans is unnecessarily high, while the energy consumption of Third World residents is untenably low. Energy consumption in the Third World must increase in order to support basic needs for survival, and development requires even more. Energy is required to transport and deliver food supplies, to refrigerate medicines, to light operating rooms, and to fire cement and brick kilns. The images of drought and famine in Ethiopia or Somalia or the floods in Bangladesh have struck many people's hearts. Tons of relief supplies pour out to disaster areas, but the lack of energy and infrastructure often leaves food rotting on the docks.

Are people in most nonindustrialized countries doomed to the restrictions placed on their quality of life by limited energy resources? Will they never be able to import adequate quantities of energy to fuel development, because they lack the energy to develop a strong, industrialized economy? Will they be the beneficiaries of new, clean, efficient energy resources that the industrialized world has spurned? Are there other possibilities for the half of humanity that depends on dwindling firewood supplies to meet their most basic energy needs? The answers to these questions have practical and ethical import for all of us.

TRACING THE PATH OF ENERGY

This book follows each energy source through its entire production cycle. A clear picture of the true impact that various resource uses have can be achieved in no other way. It is a common oversimplification to compare energy resources by focusing exclusively on the impact of a single energy source in only one phase of its production. For example, the transoceanic transport of petroleum incurs the hazard of significant damage to the local ecosystem, underground coal mining creates an ominous health hazard for the miners, nuclear power production carries the risk of catastrophic failure and radiation emissions, solar and wind energy conversion processes are expensive and often unavailable. These are true statements, but they offer at best a partial view of the costs we incur when we use petroleum, coal, nuclear, wind or solar energy, and they offer virtually no coherent view of the real choices to be made among them.

In this book, the reader is provided sufficient information and analytical frameworks to understand energy and to make informed analyses and choices. Since choices must be made, it is important to examine all alternatives carefully. It's easy to condemn one resource but pointless unless a viable alternative is proposed. The costs of each step in obtaining each energy source is presented.

This permits the reader to aggregate costs and benefits when comparing each resource and its potential applications with other resources and their applications. As with most problem solving, the first steps must be to gather information and to define and to understand the problem clearly. The need for consistent definitions is clear. Not only is it helpful to understand the differences between resource bases and reserves, but also to know what is really meant by *renewable, economic*, and *greenhouse gas.*

Because energy holds a profound place in human survival and development, and because the true costs of energy consumption may go far beyond the prices consumers actually pay, energy is a topic that can evoke strong emotions. Broadly disparate camps advocate for equally disparate approaches to energy issues. Some maintain that the economic benefits to society produced by energy's use should be expanded. Others maintain that the limitations of resources dictate limits on growth. Still others maintain that the environmental costs of the current energy resources demand a transition to new resources. Truth does not reside exclusively in any one camp. Therefore, in an effort to present a balanced work, the author will attempt to draw important points from each perspective and challenge each as well.

ENDNOTES

1. Ward, Barbara and Dubos, Rene 1972, *Only One Earth,* Norton & Co., NY, p. 10.
2. Perlin, John 1991, *A Forest Journey,* Harvard University Press, Cambridge, MA, numerous citations.
3. Smil, Valclav 1987, *Energy, Food, Environment,* Oxford University Press, NY, pp. 23–30.
4. World Commission on Environment and Development 1985, *Our Common Future,* Oxford University Press, pp. 190–202.
5. Ibid, pp. 189–192.
6. Ibid, p. 169.

CHAPTER 1

ENERGY SOURCES, RESOURCES, AND RESERVES

DEFINITIONS

Energy is that which empowers matter. Energy is such a fundamental aspect of nature that it permeates all that we perceive, so, in a sense, it is difficult to say what is not energy or an energy source. Before the middle of the present century, most scientists held the principle that energy, like matter, could neither be created nor destroyed. Einstein's Special Theory of Relativity (demonstrated emphatically by the Manhattan Project's atom bombs in 1945) caused this principle to be revised, acknowledging that matter (or mass) can be destroyed with the release of tremendous amounts of energy. The law is now expressed as the conservation of mass and energy which can be converted from one to the other. In essence then, everything is energy.

Energy is often defined as "the capacity to do work." The mathematical units of work and energy for the two are identical, and work is often the observable result of energy's interaction with matter. Energy may exist in flux or may be manifested in its interaction with matter. Energy in flux is in its purest form, typically electromagnetic radiation (such as light and heat). In its material manifestations, energy is either kinetic, potential, chemical, or nuclear. Kinetic energy is the energy of matter in motion, which can be seen in the motion of molecules as well as in a speeding bullet. This is the basis for wind power. Potential energy normally refers to a manifestation of the most enigmatic form of energy—gravity. Energy is stored by moving mass away from the dominant gravitational center and can be released by allowing the mass to fall back towards the center. This is the basis for conventional hydroelectric power. Chemical energy is stored in the bonds of molecules. It can be released by breaking those bonds and recombining the elements in less energetic bonds. This is the basis for all combustion fuels. Nuclear is that form of energy which exists as mass. Only a tiny amount of mass is destroyed in releasing vast amounts of nuclear energy. This is the basis for nuclear power.

An energy source is something from which energy originates or can be derived in a form which can be used by humans. Fossil fuels, solar energy, and fissionable uranium-235 are counted as energy sources, because science has some knowledge of how to tap them; starlight, the oxidation of iron, or the nuclear energy of non-fissionable lead are not referred to as energy sources because prevailing scientific thought sees no practical potential for converting these forms of energy into useful power for humans on earth.

An energy resource is a body of energy available to be tapped. Although the terms *energy source* and *energy resource* can be used as synonyms, the term *resource* most often refers to a quantity, such as *resource base.* The resource base of an energy source is the total amount of that energy form believed to exist. It can be defined within geographical boundaries but carries no limitations either of economy or of proof by exploration. For instance, one can speak of the United States' resource base of wind power, which would be a huge number. The term resource base does not, however, consider economic or practical limitations of production. Resource base figures then are extremely speculative. This is especially true for mineral resources, since estimates of undiscovered quantities are included.

Known resources are a subset of the resource base. Assumptions about undiscovered quantities are excluded from this term. To include economic constraints and technical feasibility of production, the term reserves is employed. Reserves make up the portion of the known resource base which experts believe has already been discovered and will ultimately be produced, under existing technologic and economic constraints. While reserves may seem to be straightforward, reliable estimates, numerous assumptions and room for error still exist in these figures.

RESOURCES AND RESERVES

There are wide variations in resource and reserve terminology. When comparing the potential of various energy resources, another complication arises because there is little consistency in the terms used to quantify energy. Oil reserves are represented in terms of barrels (millions or billions). A barrel of oil consists of 42 gallons. The energy content of oil is somewhat variable, though it averages 5.8 million Btus per barrel. Oil's production and consumption rates are normally expressed in barrels of oil per day (BOPD). Natural gas quantities are represented in terms of thousand cubic feet (MCF). Sometimes an *S* is inserted to denote that the gas volume was measured at standard temperature and pressure conditions (MSCF). The energy content of gas is consistently about 1,000 Btus per cubic foot, or 1 million Btus per MCF. (Larger volumes of gas can be reported in millions, billions, or trillions of cubic feet—but always multiples of 1,000, except for utility companies which commonly use the smaller unit, 100s of cubic feet, CCF.) The metric system uses thousands of cubic meters rather than cubic feet.

Several common terms for expressing energy quantities are shown below:

1 Btu (British thermal unit) = 1055 joules = 252 calories = .000293 Kwh (kilowatt hours)

To accommodate the wide range of values, numerous prefixes are used:

1 megajoule = 1 million or 10^6 joules

1 TJ (terajoule) = 10^{12} joules

1 PJ (petajoule) = 10^{15} joules

1 EJ (exajoule) = 10^{18} joules

1 quad = 1 quadrillion or 10^{15} Btus = 1.055 EJ

In spite of the variable energy content of oil and coal, many authors quantify energy in terms of these resources:

1 tce (ton of coal equivalent) = 27.78 million Btus

1 toe (ton of oil equivalent) = 40 million Btus

The conversions shown here are approximate averages, as there is variability even within the definition of single terms. There are, for instance, several different kinds of Btus, defined at different pressure and temperature conditions. Fortunately, this variance tends to be small (for instance, 0.1% between two kinds of Btus). The likely errors in resource and reserve estimates typically make this variance insignificant.

There are a plethora of terms applied to describe the quantities of different energy sources and the potential for harnessing those sources. The term *resource base* is perhaps the most standard. It refers to the total quantity of the given energy source believed to exist, within a given context. The global resource base of solar energy refers to the total amount of energy reaching the earth from the sun. Since solar energy arrives continuously, it must be expressed on an annual or other time frame. The African petroleum resource base would represent an estimate of how much oil exists under the surface of the entire continent. The estimate would vary dramatically from one expert to another, especially since relatively little of the African continent has been explored for oil and gas.

All of the energy sources extracted from within the earth must be found by subsurface exploration. Therefore, the known resource base of a fossil fuel or nuclear or geothermal may be a very small portion of the total. The more extensively a region is explored, the greater is the fraction of the total resource base known. But there are degrees of knowing how much of a subsurface resource exists. If exploration in a geologic region has shown that the resource is present in that region, geologists can place numbers on the amount of the energy source they might expect to find in the unexplored regions.

Ultimately, companies and policymakers alike are concerned with how much can be recovered and put to use. It is never possible to tap all of any energy resource. There are both economic and technologic constraints on the amount of energy recoverable. The term *reserves* is employed to denote the total amount believed recoverable under prevailing economic and technologic constraints. When the reserves term is used alone, it most likely refers to demonstrated or proved reserves (by production). However, there is a subcategory of *inferred* reserves used to depict the quantities that have been identified by some measurements but not proven by production. Since it is of more interest to assess the quantities of recoverable energy than of total energy in place, common recovery percentages can be applied to resource base estimates to quantify undiscovered reserves. The sum of discovered and undiscovered reserves also can be called recoverable resources.

For the energy forms in which humans tap into continuous energy flows (solar, wind, and hydropower), there are no reserves to drain. Therefore, the amount potentially recoverable (comparable to demonstrated reserves) is called the production capacity for nondepletable sources.

The amount already tapped is referred to as the *produced* quantity for depletable sources and is no longer available. The comparable term for nondepletable sources is the production and remains available for continued use. The *produced* or *production* categories are the only ones that can be measured directly for the subsurface sources and is the most certain number for nondepletable sources as well. Many other subdivisions exist in different systems to account for levels of certainty and economic producibility, making considerable room for confusion in communication about energy quantities.

Biomass and geothermal sources fall between the depletable and nondepletable categories. The reason is that both sources have significant renewal fluxes and stores from which energy can be drawn.

Unfortunately, many reports in the popular press, and even some technical writers, use resources and reserves interchangeably. For example, later chapters will show that the ultimate recovery from a known petroleum resource is not likely to exceed 30%. Thus, interchanging even the terms *known resource* and *reserves* incurs a 300% error. If *total resource base* and *reserve* terms are misused, the error is likely to be as much as several thousand of percents.

This is a significant part of the reason why people were confused about "running out of oil." Even if resources and reserves figures were not misused, the popularity of referring to the number of years of productive life remaining for current reserves is problematic. There is no physical significance in referring to the length of time that current reserves will last under current production rates. New discoveries constantly add to reserves, while production subtracts from them. Using reserve-to-production ratios to cite how much oil is left ignores new reserve discoveries and the fact that improvements in technology and increasing price (which would be expected to accompany any supply shortage) increases the ultimate recovery, and thus the portion of known resources occupied by reserves. The quotation "At the present rate of consumption [the total oil reserves in the United States] would be consumed in 14 years . . . the American supply of oil will be completely exhausted in a quarter of a century" may appear to be lifted from headlines of the 1970s.[1] Actually, this quote is from a

1919 study of American fuels; it exemplifies the shortcomings of comparing reserves to production rates.

There are some good reasons for the lack of standard terminology. One reason for differences is that the renewable resources are different than the finite resources. For instance, it really makes no sense to speak of the ultimate recovery of solar energy, because of the essential need to predict how long humankind will survive! Although the term *resource base* can be used for any energy form, some extra care must be exercised in its use with the renewables. Four million quads (quadrillion Btus) of solar energy reach the earth annually. This is nearly 15,000 times the world's (commercial) energy consumption in 1980.[2] So, would it be reasonable to say that the resource base of solar energy is 15,000 times the world's energy demand? Photons (units of light energy) captured for the generation of electricity are not available for other uses such as photosynthesis, evaporating water to form clouds, or sunbathing. Since a fair portion of this energy is necessary to sustain life on earth, even the most optimistic estimate of the solar resource base must be seen as a relatively small fraction of the light incident on the earth's surface. Nevertheless, solar energy has an abundant resource base, which is at least theoretically capable of meeting all of humanity's energy needs. Solar energy, and some other renewable energy resources, are limited by the technologic capability to use and store the energy rather than by their overall abundance.

Consider the ways in which energy resources are typically classified: renewable energy, fossil fuels, and nuclear energy sources. In fact, these categories can be very misleading. Consider first what the terms mean, literally, and then how they are actually used.

RENEWABLE RESOURCES

First, renewable resources generally include solar, hydroelectric, wind, and biomass. (Biomass refers to organic matter which was alive relatively recently and can be used as a combustion fuel.) Solar energy is *not* literally renewable. The sun is burning itself out by fusing hydrogen atoms into helium and releasing nuclear energy as electromagnetic radiation. In fact, some 657 million tons of hydrogen fuse to become 653 million tons of helium every second in the sun: The missing four million tons represent the mass converted into energy every second.[3] There is no regenerative process at work; once a pair of hydrogen nuclei fuse, that's it, they don't split apart and fuse again. The sun also is not large or hot enough to carry the fusion to the next step. Solar energy is called renewable simply because it keeps coming to earth, through no human intervention, and will probably continue to do so for the rest of human history. Although it is a remarkably vast resource, the sun is finite and depleting; scientists estimate that it will produce energy for about 5 billion more years.

Some authors claim that all other renewable energy sources are derived from solar energy. This is not precisely true in that winds are partially a function of the earth's rotation, and tides (which this text considers as a type of hydroelectric energy source) are a function of the gravitational attraction of the moon. Certainly, though, biomass and river hydroelectric applications utilize naturally stored energy derived from the sun. So, while the growing process and hydrologic cycle are constantly recurring processes, they are dependent on an ultimately finite resource.

Geothermal energy, which many people do not include in the renewable category, may be quite renewable; a fairly constant heat flux from the earth's interior is renewed by the heat of natural radioactive decay, gravity, and perhaps other forces. There is some debate as to the relative significance of these processes. Like solar energy though, geothermal energy represents a resource that ultimately is very large and has a renewal process.

All of the preceding is academic and even trivial in a sense. Certainly the immensity of the solar energy resource base cannot be questioned, nor can the fact that human use does not deplete the sun. What is important is that renewability is not an absolute. For example, although fossil fuels are not generally considered renewable resources, they are nonetheless renewing from decaying biomass at a rate of approximately 16,000 tons of oil equivalent (toe) per year.[4] (The rate was much higher during the times in geologic history of a warmer global climate and vast swamps and is lower now due to human activity—draining swamps, felling forests, and generally taking over much of the land, but renewal certainly has not stopped.)

Most of the resources actually called renewable (like wind, solar, and hydro) are produced from sources which human activity does not affect. Biomass is the most important of the *so-called* renewables to question. Like coal, oil, and gas, biomass forms continually, and it renews at a much higher rate than the fossil fuels. However, though it is constantly being produced, new biomass resources depend on the existing reserve base (forests, etc.). Human activity can and does deplete the resource base, which in turn decreases the renewal rate.

FOSSIL FUELS

What are fossil fuels? Basically, they are the same as biomass. The difference is simply a matter of how long the material has been dead. Fossil fuels are (according to most prevalent geologic theory) the carbon-laden remains of dead organisms which have been altered to some extent by temperature, pressure, chemical reaction, and time. They include coal, oil, gas, and kerogen (oil shale), which are constantly being renewed. The same processes of death, decay, natural burial (under ocean and lake sediments), and geologic metamorphosis are at work today as they were millions of years ago, when current oil, gas, and coal fields were forming. In fact, one can travel to the Okefenoke swamps in Florida or Dismal Swamp in North Carolina and observe the early processes of making a future coalbed. The fossil fuels are deemed to be nonrenewable because their renewal rates were exceeded by their consumption rates almost as soon as people started using them for fuel.

NUCLEAR ENERGY

Nuclear energy is released in large quantities when small quantities of matter are destroyed. This occurs either when large nuclei split to form smaller ones or when small nuclei come together to form larger ones. Splitting large nuclei is the easier process for humans to initiate because some naturally occurring massive nuclei, such as those of uranium 235 are unstable. If humans simply cram another neutron into the unstable nucleus, it will split, releasing a large amount of energy. Even larger amounts of energy are released when two nuclei are brought together in fusion, but forces such as the like electrical charges of nuclei make it very difficult to bring nuclei close enough to fuse. A tremendous amount of energy is required as a *trigger* to initiate a fusion reaction. It is this obstacle that stands between harnessing controlled nuclear fusion and existing technology. The potential of fusion can be seen clearly on any sunny day. Solar energy is derived from fusion energy in the sun.

ABUNDANCE AND AVAILABILITY

Abundance and availability are two distinct issues. Many potential energy sources have abundant resource bases, but technologic, geographic, or economic factors keep them from being available for practical use. Economic analyses often predict energy costs which would allow various new energy sources to achieve significant market penetration. These predictions have produced considerable confusion and have probably been one contributor to a common distrust of *so-called* energy experts by the general populace. For many years, various experts have predicted that solar energy, oil shale, coal gasification, etc., would become competitive when the price of energy increased by a few percent. Energy prices have increased, but where are the major solar and oil shale plants?

The complete answer to this question is not simple because

there are different questions and answers for each resource. The simplest part of an answer is that real energy prices have not increased significantly. In fact, energy prices have barely kept pace with long-term inflation. In the consumptive United States, the price of natural gas was held constant by federal mandate for the 20 years following the Phillips Decision of 1954, which froze the price of natural gas. During this time oil, gas, and coal prices remained stagnant. Other prices did not. When energy prices skyrocketed after the 1973 embargo and the emergence of OPEC dominance, they were largely just making up lost ground on the rest of the economy. The era of short supply in the market place did cause energy prices to overshoot a stable point, and they fell precipitously in 1981 and again in 1985. What happened? To some extent, many energy industry analysts asked the same question.

Generally speaking, the market place responded in two logical ways to the perceived supply shortage. On the one hand, consumers began trying to conserve diminishing (and increasingly costly) energy resources. Governments mandated manufacturers to make more energy-efficient appliances and increased tax incentives to consumers that conserved. In fact, some of the energy utility companies (gas and electric) began to recognize the advantage of facilitating conservation loan programs, and sometimes even rebates, to consumers for energy-saving efforts. At the same time, the energy-producing companies responded predictably to the increasing price by investing more to produce more. Wells were drilled deeper and farther offshore, advances in technology increased the recovery of hydrocarbons from existing wells dramatically, and some new energy sources even began to penetrate the market. The effects of these responses to the perceived shortage and increasing prices hit with cumulative effect at the end of 1981. Oil prices fell and continued a slide through much of the 1980s. Since oil dominates the industrialized worlds' energy markets, its price leads other energy prices.

Was the shortage of the 1970s a mere price-manipulative ruse? No. *Shortage* is not a sufficiently precise term to analyze the energy market. What the United States experienced may be referred to as a *strategic supply shortage:* the United States was (and is) a major net energy importer. The same is true of other industrialized powers such as Japan and Germany. A nation's economic strength is clearly weakened by its dependence on essential imports. The United States experienced a strong sense of vulnerability during the two oil crises of the 1970s. Actually, while these events were very real supply disruptions which brought a shocked awareness of the finitude of hydrocarbon resources, they did not represent a global resource shortage.

The United States built a world-dominant industrial society partly on innovation and partly on an abundance of cheap energy. Some scholars may trace this building of American wealth on cheap energy supplies back to the 17th and 18th century slave trades. Leaving that argument to the social scientists, consider briefly the growth of American industrialism with local energy production. By the 19th century, New England whalers were producing whale oil and Appalachian coal miners were providing an abundant supply of energy sources to the newly independent nation on which (along with other abundant raw materials) it built a strong industrial base. When the American Civil War broke out, the South's wealth was largely attributable to forced labor of other humans in agriculture, while the North shifted its base of wealth from mercantile trade to manufacturing. The industrial strength of the North was a significant factor in the outcome of that war. By the early decades of the 20th century, the United States had become an industrial giant and the world's largest energy producer.

COMBUSTION FUELS

COAL

Coal is a resource which will certainly not be exhausted during the lifetime of anyone living in the 20th century. The known resource base existing in the last decade of the century would continue to meet current demand for as much as 400 years. The proved reserves estimated by the World Energy Commission in 1980 would last nearly 80 years, at a continued 2.6% per year growth in consumption.[5] If an abundant energy supply is the priority, coal must not be overlooked. If environmental preservation is the priority, coal resources must be developed very carefully. The latter concern is pertinent to this section because converting a resource into a reserve is a function of cost and technical capability. Any decision to make primary energy sources bear an environmental cost will necessarily work against the expansion of coal reserves.

Occurrence • Coal is very interesting because we can trace its entire evolution without interruption. A number of conditions permit the formation of coal, but it is very often formed from the remains of plant matter that decayed at the bottom of a swamp. As trees and other swamp plants die, they tend to fall into the surrounding water. Although the dead plants are naturally devoured by creatures such as termites and oxidized in the air, the water burial slows these processes, so that not all of the dead material is consumed by either process. Over millennia, a great deal of biomass accumulates, the older material being compacted by the weight of overlying vegetable matter. These accumulations are called peat bogs. Thus, peat can be material that has died quite recently—a biomass fuel. When a peat bog is buried by geologic processes under rock sediments, the coalification process begins.

During a swamp's evolution, rising ocean levels inundate the swamp, and falling sealevels drain the swamps. Geologists refer to rising and falling sea levels as *transgressing* and *regressing*. Swamps tend to move as the areas of stagnant water shift, and vegetation follows in this rather cyclic and very slow process. Paleoecologists study this evolutionary dance, with land and water leading and vegetation following, as the results of this process can be seen in the distribution of many layers of coal in a region.

The formation and evolution of coal involves compaction,

which drives out fluids. A large portion of the fluid driven out is water; so as the coal is compacted, more of the solid, carbon-rich material is left in a given volume or mass. The carbon compounds are the source of coal's chemical energy; so the compaction increases the amount of energy contained in a given volume of coal.

Rank and Quality • Older, more compacted coals generally have higher chemical energy densities (i.e., more energy per unit weight) than younger coals. The higher energy density makes more mature coals much easier to transport, since less coal is required to produce a given amount of energy. The maturity of a coal is described as its rank. The Appendix shows the various ranks of coal and properties commonly associated with those ranks. Lignite is the youngest of the coals; it represents peats that have been buried under sediments (often after a sea transgression). The next step in the evolution of a coal is bituminous. There are many gradations within the general bituminous rank, reflecting decreasing moisture and volatile matter (like methane) as the coal has been subjected to deeper, longer burial. Anthracite is the oldest coal and the most altered from its original vegetal state. Virtually no moisture or volatile matter remains in this rank. With increasing rank, the energy content increases, along with the solidness of the material. The simplest classification scheme differentiates only between *hard coals,* with energy contents of 10,260 Btu/lb or more and *soft coals* with lower heating values. The gain in energy content is offset somewhat by the fact that some of the fluid driven off during compaction is volatile matter, which is generally combustible methane. Some people are initially surprised to see that ash content has a tendency to increase with rank. Most of the material driven off during compaction is water; the ash concentration (on either a volume or weight basis) increases along with the carbon content. Does this make high rank coals dirtier to burn? No, the ash content must be compared to the energy content to evaluate this question. Rank is determined by assay. Interestingly, a preferred test for rank has nothing to do directly with its heating value or depth of burial. Rather, coal is ranked by how shiny it is. The geologists and geochemists, of course, have devised a more impressive sounding name for the test: *vitrinite reflectance.* But it means simply that the vitrinite component (mostly carbon) is shiny and becomes more concentrated with age and rank. Those of us who have been amateur rock collectors were quick to notice that anthracite is almost as shiny as obsidian, while bituminous is a dull black. Lignite is often called brown coal for its lackluster appearance, and peat looks like what you might draw from the bottom of your composter.

A number of other assay tests, including agglomerating properties and free swelling index, relate to the coal's behavior during slow and rapid oxidation. These properties can indicate for what use the coal is best suited.

Coal has an occasionally touted potential for conversion into fluid fuels—*synfuels*—through liquefaction or gasification. Synfuels overcome the definite convenience barrier and to some extent the environmental barrier; contaminants, such as sulfurous compounds, can be removed in the synthesis process. There is, however, an efficiency price as these processes consume energy. They have continued to add sufficient cost to the coal product as to make synfuels generally unable to compete with oil and gas thus far.

Table 1.1 shows known resources and proved reserves for each region of the world, both in terms of tons of coal and of Btu equivalence. It should be noted in viewing this table that many of the regions showing very little resource are also grossly underexplored. For example, the United States Geological Survey has been conducting research into the possible presence of significant yet hitherto unknown low grade coal resources throughout much of sub-Saharan Africa.

Table 1.1 Hard Coal Proven Reserves by Region

Region	Billion Tons	Quads
Former Soviet Union	4,860	135,100
USA	2,570	71,400
China	1,438	40,000
Western and Central Europe	600	18,300
Australia	262	7,300
Sub-Saharan Africa	169	4,700
Canada	115	3,200
Southern Asia	62	1,700
Latin America	29	800
East Asia	11	300
Total	10,116	282,800

Source: Hedley, Don 1986, World Energy: *The Facts and The Future,* Euromonitor Publications, London, p. 186.

In addition to comparing these numbers to the roughly 300 quads of energy consumed annually in the world, the reserve distribution is significant. Most of the proved reserves are concentrated in industrialized countries. This fact is magnified when the reserves in and controlled by South Africa are seen to account for more than a third of all sub-Saharan Africa's reserves. Does this mean that only regions containing coal resources have industrialized, or does it mean that reserves are only identified in industrialized countries?

OIL AND GAS

The global known resource base for oil has been estimated at 2.2 trillion barrels, which (at 5.8 million Btus per barrel) equates to 12,700 quads. Of this, 610 billion barrels were considered to be recoverable reserves.[6] The work of M. King Hubbard, in the 1960s, would suggest that undiscovered resources are likely to add about another 600 billion barrels of reserves—roughly doubling the probable reserves which experts expect to recover from the resource base currently known.[7]

Occurrence • Oil and gas, like coal, are commonly considered to have been formed from the decayed remains of huge quantities of dead plants and creatures. The conditions under which the matter was buried protected the remains from consumption by other living things or oxidation. At the risk of offending advertising agencies, though, scientific theory does not suggest that dinosaur remains are significant constituents of petroleum deposits. Although some dinosaurs were massive creatures, even a very modest petroleum reservoir would have required that millions of the large fellows die in the same place. More likely, the teeming life of smaller aquatic creatures feeding on dense nutrients spilling into the oceans (or sometimes lakes) constitute the deposits. Consequently, many petroleum reservoirs are believed to be in regions of paleo-deltas, where rivers carried massive nutrient and debris streams into the ocean.

As time passes, the rock sediments (mud, silt, and sand) falling out of the rivers pile up, forming massive beds of sand, silt, or clay—or of chemical precipitates such as calcium carbonate. As with coal, more rock material is deposited on top of the rich beds of decaying organic material. Under *normal* circumstances, as rocks are deposited, the fluids occupying pore space in the rocks form a continuous phase, so that the fluids at any depth support the weight (hydrostatic head) of the overlying fluids. Similarly, the rock grains or particles are in continuous contact, supporting the load of overlying rock grains. This normal condition is a result of a sedimentation and compaction process in which the weight of additional sediments drives fluids from the lower sediments being compacted. However, as depth of burial increases, it becomes increasingly likely that impermeable beds prohibit adequate expulsion of fluid to maintain a continuous system. In this case, the trapped fluids must carry some of the rock load, causing the pressure to increase more rapidly. This condition is called *overpressure* and is typical (if not *normal*) in sediments at depths of 10,000 feet or more.

Three conditions are required for an oil or gas reservoir to exist: a porous and permeable rock, a trap, and a source rock. The porous and permeable rock constitutes the reservoir itself. Oil and gas are contained in the vast number of tiny pore spaces between rock grains and crystals in beds of rock such as sandstone or limestone. As burial under increasing rock loads forces fluids out of the compacting shale material, the lighter oil and gas are forced up through the water contained in the porous and permeable reservoir beds. The oil and gas then occupy the tiny void spaces between the grains or crystals of the reservoir rock. Some of the water in which the rocks were deposited always remains, though. Generally, the water coats the rock grains while the oil and gas fill the middles of the pore spaces. (A microscopic view of the pore space of an oil reservoir is depicted in Figure 1-1.) Lighter hydrocarbons migrate upward until they reach the surface and escape to the atmosphere, unless they encounter the trap, which is an impermeable obstacle.

Almost all oil reservoirs contain some amount of gas. Some of the gas can be dissolved in the oil, like carbonation in a soft drink; but if more gas is present than can be dissolved, it forms a free gas cap. Since gas is very light and mobile, small amounts of the gas can escape through very tiny fissures or porous paths to the surface. The gas which escapes to the surface is predominantly methane.

Fluid hydrocarbons virtually always occur in aquatic (marine, fluvial, and lacustrine) sediments, where water originally filled the pore space of all reservoir rocks. Even when vast amounts of oil or gas migrate to fill a reservoir, they cannot displace all of the original (connate) water. If huge amounts of hydrocarbons do not move into a reservoir, more water may be left behind. The more water sharing the pore space with the oil and gas, the less desirable is the reservoir.

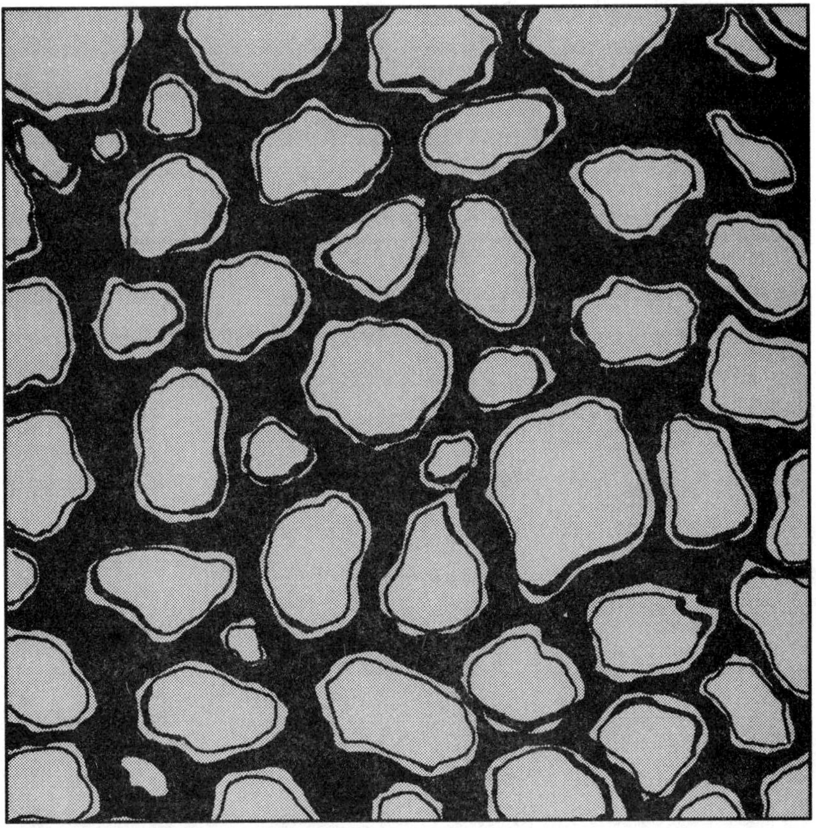

Figure 1–1 Oil reservoir pore space.
The light areas represent sand grains, with light shading representing the water coating and dark shading indicating the oil or gas in the middle of the pore spaces. This would be a good oil reservoir rock. Sketch by Mr. Loyd Brown.

Oil Markets • Through the middle of the 19th century, petroleum (which had been discovered seeping to the surface at such places as Oil Creek Pennsylvania) received intermittent and sometimes flamboyant attention. Although it was combustible, it was not found in a suitable abundance on the earth's surface to make it a commercially attractive fuel, and it produced a disagreeable smoke. Some enterprising entrepreneurs skimmed it from the creeks to bottle and sell as miracle medicines. Around 1850 Samuel Kier built the first petroleum still for oil produced from salt wells. The distilled product was an excellent lighting oil. A sudden increase in demand caused prices to jump from seventy-five cents per gallon to two dollars per gallon during that decade.[8]

In 1859, "Col." Edwin Drake hired a salt driller to drill a well near Titusville, Pennsylvania. The cable hit a fissure from 62 to 67 feet of depth, and about 10 barrels of oil per day began to flow. This was an abundant production rate for a product which had previously only been sopped up from natural seeps on the ground or floating on the surfaces of streams. The newly abundant petroleum resource quickly showed its fuel potential and spawned the first drilling boom. By the turn of the century, hundreds of wells had been drilled in western Pennsylvania and northern West Virginia. The spurt of drilling sent the price plummeting. Although demand did grow rapidly, it lagged behind production. Soon production caught up, but continued to fluctuate dramatically as new discoveries drove prices down and increasing demand drove it up.

Liquid fuels have clear advantages over solid fuels. Liquids can flow from the storage reservoir to the point of use, like kerosene moving up a lamp wick or gasoline pumped into an engine. The flow of minute quantities of fuel to where it is needed provides for uniform and readily controlled combustion. Furthermore, the fluid fuels can be processed (refined) through simple distillation processes to meet a variety of combustion criteria, including a clean-burning criterion. These advantages drove petroleum's ascendancy in consumer preference, so that it soon surpassed coal in the market place.

The petroleum industry's early growth was rapid, if "crude." Before long, petroleum geologists abandoned the notion that oil flowed in subterranean rivers beneath surface rivers. Improvements in drilling and recovery technology followed. By early in the 20th century, oil was produced from shallow waters offshore in the Gulf of Mexico and in various locales around the world.

Never before had an energy resource been discovered which, once found, could simply flow out of the ground to the consumer, at whatever rate the natural subsurface pressure could sustain. The largest portion of the investment was in drilling the well. Once drilled, the on-going extraction was neither labor intensive nor particularly expensive. Thus, it was the ability to produce great volumes at low incremental cost that made petroleum a source of wealth. Although called black gold, its value per unit volume or weight has never been very high. Even in the soaring prices of the 1970s, oil (or its refined product gasoline) hardly exceeded the price of other liquid products, such as milk. But a single oil well may produce as much as 20,000 or even 30,000 barrels per day initially. Even at the lowest point the depressed prices of the 1980s reached, such a well would generate revenues of approximately a quarter million dollars a day. There is no doubt that such wells were always very lucrative. Such prolific discoveries, however, are extremely rare. The probability of an exploratory well being commercially successful can be as poor as one in 20 (depending on the extent of geologic information available). While no cow produces 30,000 barrels of milk per day, a farmer does not have to buy 19 cows to find one that produces, nor does he have to spend $12 million on a single cow which is likely not to produce.

The liquid fuel was more convenient than any other and of greater energy density. It made development of self-contained horseless carriages possible, which soon became so popular that most adult Americans owned one. Oil field discoveries continued to be made throughout the United States during the early 1900s; meanwhile, U. S. and European-based firms established production around the world, including the super-giant Ghawar field in Saudi Arabia.

The abundance of energy was good for the growing American economy. Growth in demand was good for the petroleum industry, which relied heavily on high volume sales to make its profits. As international production met a larger share of market demand, the governments of the exporting countries realized that a great source of national wealth was being produced by foreign companies for foreign consumers. The nationalization of oil production followed this realization. In some cases, the process was relatively amicable, as in the case of Saudi Arabia's working participation in ARAMCO (the Arabian American Co.), with the original producing companies maintaining working interests in the reorganized company. Sometimes it was not, as in the case of Libya's nationalization of British Petroleum's interests and those experienced in Iraq and Iran.[9] In 1960, Venezuela, Saudi Arabia, Kuwait, Iraq, Iran, and Indonesia formed the Organization of Petroleum Exporting Countries (OPEC). It was not until 1973, though, that they felt sufficiently powerful to insist on a price increase to the western-based petroleum companies, and even walk out on negotiations. At the same time, the newly perceived power was used by the Arab nations as a political weapon. By this time, the United States was dependent on imports for over 36% of its oil supply and was shaken deeply by an embargo.

American people experienced the shortage in long lines and increasing prices at the gasoline pumps. OPEC members realized that their oil was much more valuable to them in the long term if sold at lower production rates, but at higher prices. Again, to draw the agricultural contrast, a farmer does not have the option of postponing production; a crop not produced today is forgone production. OPEC's oil not produced in 1973, though, remained available to be produced in the future. Some people cry "foul" at this form of market manipulation, while others cry "foul" at the monopsonistic manipulation of the consuming Americans. (Monopsony is the converse of monopoly; a single consumer, or set of consumers acting in unison, control such a large share of demand that they are able to control the price.) The question of who the villains are, if villains at all, depends largely on one's perspective of nationalism, free market, and justice—issues for which there are no simple answers. For now, it is important simply to understand the basis of the supply shortage.

Price-shocked Americans, suddenly seeing the cheap, abundant energy supply curtailed, faced dramatic continuing price increases and accomplished nearly a 20% oil conservation and an

11% reduction in total energy use over the 1973 to 1983 decade. (Actually, the reduction occurred from a peak energy consumption in 1979 to a minimum just four years later in 1983. The change required reversing a trend of growing energy use and turning it around.)[10] At the same time, a fervor of activity was spawned in domestic American production, which had lagged for years. Relatively shallow giant fields had been discovered decades earlier and were being depleted. While foreign oil flowed from giant and supergiant fields, there was little incentive to drill deeper in search of new fields or to employ costly new recovery technologies in older fields. (It is important to note that there is no proportional increase in expenditure to produce wells up to their limit—a limit which is still very high for many international wells, while most American wells have already produced at their limits for years.) The domestic shortage was real in two ways: domestic production and the domestic reserve base were falling, while increasing demand had been met by rapidly increasing oil imports.

American petroleum engineers and chemists were well aware that the typical oil field produced to depletion left between two-thirds and three-fourths of the original oil in the ground. (The technical reasons for this will be discussed in the next chapter.) Reserves represent the portion of a known resource believed to be producible under existing technologic and economic constraints. So, petroleum researchers went to work vigorously on *enhanced recovery* technologies, which had the technical potential to double American reserves, simply by permitting production of a larger portion of the oil in known fields.

The 1970s did not generate the first oil crisis, and certainly not the first energy crisis. Energy crises of one kind or another have spurred the transition from one energy source to another and even influenced the rise and fall of nations. Earlier oil shortages in the United States and Europe were given the crisis label in both world wars. In December of 1943, the United States Secretary of the Interior (and Oil Czar) Harold Ickes, published an article entitled, *We're Running Out of Oil.* In 1948, the United States declared an "Energy Crisis." The Suez Crisis of 1956 produced serious energy-market disruptions, with an embargo of petroleum against Great Britain and France but not the United States. Arabian producing states issued an embargo after the Six-day War in 1967, as well as the more famous and successful embargo of 1973, which drove prices up dramatically.[11]

With rising oil prices, western financial institutions were suddenly eager to extend credit to the oil-producing Third World countries. Traditional development theories purport that a massive injection of capital is the requisite for development. (Ahmad Abubakar discusses why a lack of capital-absorbing infrastructure in a lesser developed country can be a critical failing to the traditional theory.)[12] Many poor, oil-producing countries took massive loans, with their oil wealth as collateral.

Demand leveled (and even decreased) while technology added reserves as did a boom in worldwide exploration, causing supply to catch up with demand. Countries with large loans were under pressure to maintain high production levels. The price of crude oil slipped on the world market even as production capacity was still increasing. The option to restrict production had been bartered for international credit by too many oil-producing countries. OPEC tried to establish production limits for each member country. But several non-OPEC exporting countries had emerged, and several debt-burdened member countries opted to cheat on production quotas in order to meet their debt-repayment schedules.

With the full impact of these market pressures, the price of oil plummeted within a few years to less than one-third of the maximum price it had reached. It was again cheaper to buy oil from foreign producers than to produce marginal American wells. By 1990, America was importing half of its crude oil—more than it had prior to the traumatic 1973 embargo. If import independence is deemed a priority, then a shortage exists and has continued to exist. If, on the other hand, pure free market policies are elected (including for foreign markets), a world supply shortage does not exist.

We as Americans are especially inconsistent in how we view

energy supplies. On one hand, Americans insist that the free market should exist without constraints and should buy energy from the cheapest source. On the other hand, when a supply disruption occurs and a domestic shortfall looms, we want our politicians to "keep the oil companies from raising prices." We want an abundant supply so that we can drive wherever we want whenever we want, and still want to force the foreign oil exporters to give us "our" oil, from their fields. During a period of tightened demand surrounding United States sanctions against Iraq (and Iraqi-controlled Kuwait) in 1990, a political/consumer outcry arose against the oil companies for increasing the prices on oil products, while supplies that had been produced earlier under lower prices were available. This presents an interesting irony. When prices fell in the 1980s, there was no mechanism to keep prices up on a product from very expensive wells which had been developed under the higher prices of the late 1970s. Does it make sense to expect prices to fall on production from expensive wells when it is a buyers' market, yet demand that prices not increase when a sellers' market appears?

Furthermore, the furor about increased oil prices came during Earth Year, when America and much of the world was recommitting itself to the environment, responding to fears of global warming, acid rain, and so on. The fossil fuels were widely cursed in the popular press as the main culprits in the environmental crisis. Yet the accomplishment of conservation through rising prices was seen as abhorrent.

This is not meant to suggest that the vicissitudes of fair weather environmentalists who are also gluttonous energy consumers diminish the importance of the environmental movement. Nor is it meant to suggest canonizing members of the petroleum industry. Rather, it should draw attention to the fact that while companies may reflect some of the personality of their leadership, the companies are not human. Energy companies, like any other, respond to market stimuli in efforts to maximize their profits. Accusations of shortsightedness in failure to put adequate emphasis on preserving the environment, or of failing to preserve marginal markets (poor consumers), can have weight but must be shared by all. Accusations of obscenity or malice are pointless.

Future Production Potential • The market forces controlling any company's behavior are twofold: the customers who buy their product and the customers who buy their stock and thus own the company. The latter individuals set the agenda for the company's strategic priorities, while the former set the agenda for the company's production. As long as consumers demand large amounts of energy, which has long been tied closely to economic well-being, they are establishing a force for increasing production of the energy source. As long as shareholders put a priority on the rate of return paid by the company, they are establishing prices favoring the short term.

So, with all this said, oil and gas should be viewed as finite resources which are not particularly close to ultimate depletion. At the writing of this book, supply continues to respond to demand, and it will most likely continue to do so well into the 21st century. In fact, existing world reserves would carry us into the third decade of that century at current consumption levels. Many new reserves remain to be discovered. If geologists and engineers did not believe that, exploratory drilling would effectively stop. As demand exceeds supply, substantial reserve additions remain possible within known resources by producing older, depleting wells, and employing enhanced recovery techniques.

Much of the world also remains unexplored. Consider the enormous disparity between 600,000 producing wells within the United States and only 6,000 producing wells on the African continent, in spite of Africa containing much more sedimentary basin area than the United States. Furthermore, even in fully developed oil fields, oil reserves are never likely to be more than one-third of the total amount of oil known to be in the reservoir. A great deal of oil remains to be discovered, and, if needed, a great deal more can be extracted from known fields (Table 1–2). So, in answer to

Table 1.2 Oil Reserves by Region

Region	Oil (Quads)	Gas (Quads)
North America	495.0	332.7
South and Central America	384.2	160.3
Sub-Saharan Africa	122.6	104.8
Mideast and North Africa	3903.8	1388.1
Western Europe	105.3	191.9
Asia, Pacific and Eastern Europe	600.0	1811.0

Derived from data in the *Oil & Gas Journal*, Dec. 25, 1989, pp. 44–45.

the often asked question, it seems highly doubtful that anyone alive at the writing of this book will live to see the exhaustion of oil and gas reserves.

Unconventional Oil and Gas Resources • Many authors discuss tar sands, tight gas sands, and geopressured (gas-bearing) aquifers as separate resources. In this text, they will not receive special treatment. Each of them simply represents an extreme of normal oil and gas field conditions. Tar sands are merely oil reservoirs which are unusually devoid of the lighter hydrocarbon molecules that make oil fluid. They may have been conventional oil reservoirs at some point in geologic history, but through erosion, the reservoirs may have become exposed to the surface, allowing nearly all of the lighter molecules to evaporate over the eons. Tar sands are noteworthy especially in Canada, where the Athabasca tar sands are extraordinarily extensive.

Tight gas sands are simply gas reservoirs that have high shale contents and are thus relatively impermeable, so fluids have difficulty flowing through them. Gas has such low viscosity, because its molecules are so small, that it is possible to achieve reasonable production from the tight sands. The technologies for stimulating the productivity of an oil or gas reservoir, discussed in the following chapter, can be applied to establish production in these reservoirs. The numbers given in this book for reserve and resource base estimates are meant to include unconventional resources, but tight gas sands are likely to be grossly underestimated because the shaly reservoirs in which they occur are often ignored in standard exploration.

Geopressured aquifers are formations which are essentially 100% water-saturated sandstones, but gas is dissolved in the water, like carbonation in a soft drink. All subsurface formations are technically geopressured, that is, they are under pressure of overlying earth formations. The fact that especially high pressures make it possible for water to hold appreciable amounts of gas in solution is what makes geopressured aquifers special. If prices warrant, it is possible to produce these water-bearing formations in order to extract the gas.

KEROGEN

Kerogen is, like coal, a primary energy source with a vast resource base. As it has not proven to be economically viable yet, it has virtually no reserves. Some projects, such as the UNOCAL Parachute Creek site in northwestern Colorado, have produced modest amounts of *shale oil,* but with some form of subsidy such as price guarantees. The technologies for extracting synthetic crude oil from kerogen have not succeeded in overcoming fundamental obstacles in a sustained, cost-effective manner. This resource is not likely to be tapped on a significant scale unless depletion of oil and gas resources forces energy prices back to the highest level seen during the energy crises. However, prices that high would probably facilitate the commercialization of other resources.

BIOMASS

Biomass, the chemical energy of recently living matter, is a complex primary source to assess. Some authors insist on exaggerating its resource base by ignoring the other necessary uses for what they call "agricultural waste products." Other authors ignore the current and past contributions of biomass, which was the dominant fuel for all of humanity until overtaken by coal around 1880.[13] Approximately half of humanity still relies on biomass to meet its energy needs. The fact that these typically impoverished people's energy needs per capita are so low allows many energy analysts to ignore the contribution of biomass fuels. While some experts place the total energy content of all biomass at between 15 and 20 times the amount humans currently get from commercial energy sources, "they also note that only a fraction of the total biomass has any potential to be used for energy." In order to meet the criteria of sustainability, annual biomass consumption must be limited to a number less than the annual biomass growth rate. (It must be less than the growth because biomass has many uses other than energy production, including human, livestock, and wildlife food production.) The energy content of the earth's collective biomass growth is estimated to be approximately 3,000 quadrillion Btus (quads). Of this, 23% is found in established swamps, grassland, tundra, etc.; 29% in forests; 10% in cropland; and 38% in aquatic systems.

A few logical assumptions can be employed to reduce the estimated biomass energy production to a plausibly sustainable basis. One International Institute for Applied Systems Analysis (IIASA) study suggests that a maximum of 40% of biomass growth on land could be "prudently cultivated." Of the resulting 750 quads of harvestable energy content, 56% is already harvested to produce lumber and crops. In addition to the 330 quads remaining, another 60 quads might be gleaned from waste products of the agricultural and lumber industries. Considering probable conversion efficiencies of somewhat less than 50%, only about 180 quads could be harnessed from biomass, assuming excellent management and production efforts.[14] Since the resource base does not consider recoverability or efficiency, the total resource base can be viewed as 390 quads per year.

Some might argue that improved cultivation practices might

increase the annual yields, but high yield agricultural practices are very energy intensive. In fact, additional losses in gathering and transportation need to be counted against the total when trying to arrive at a figure for biomass reserves. Furthermore, a large portion of the biomass, even in forests, consists of things other than woody biomass, including wildlife, small plants, and micro-organisms. And a total harvest implies completely stripping the land—hardly an environmentally friendly activity.

In terms of converting the still large biomass resource base into reserves, the *best use* of land and living resources must also be considered. The world's societies seem to be growing increasingly conscious of including an environmental component. The IIASA study did exclude 60% of land-based biomass as not feasible to harvest for energy, perhaps for reasons of accessibility (Arctic tundra is likely too sparsely distributed to harvest effectively) or environmental concern (not felling all of the virgin forests). Whatever the reasons, the numbers also indicate that more than half of the biomass available to harvest is already being used. It must also be recognized that gleaning all of agricultural wastes or all of a forest's growth mass is not environmentally sound either. Debris and stubble in fields or on forest floors are essential in preventing erosion and supporting the small life-forms that digest organic material to make soil.

NONCOMBUSTION SOURCES

SOLAR POWER

Solar energy, of course, has an immense resource base: of the sun's enormous energy production, about 2 billionths of it reaches the earth's surface. This minuscule fraction still amounts to 3,900,000 quads per year. After accounting for the portion which must fall to the earth for other purposes (including photosynthesis and warming the earth), 975,000 quads are normally reflected back to space.[15] This number equates to about 3,000 times humanity's annual energy consumption. Even if only a small portion of this amount could be viewed as accessible, the resource base is immense. Reserves are another story altogether. Even the most optimistic reports indicate a projected contribution of solar energy barely reaching 10% of electricity requirements in the United States by the early part of the 21st century. As long as energy prices remain relatively low in the industrialized world, even this modest market penetration is unlikely. It would probably be necessary for energy consumers to begin paying serious environmental costs for this transition to be made. (Consumers do pay environmental costs in the forms of increased costs for industry to comply with environmental regulations and in terms of increased health care costs, but the indirect costs are not seen "at the pump.") How is it possible for an urban worker to know how much they are really paying to commute into town and back to the suburbs, when some of the indirect costs may show up 30 years later, when asthma progresses to emphysema? Many environmentalists would say that even including such indirect costs does not fully represent the opportunity costs of energy consumption. How much will it ultimately cost to lose forests to acid rain and to the firewood market? What vast pharmaceutical wealth will go undiscovered in the lost forests? If this question seems remote, consider that approximately half of prescription medicines are derived from wild organisms, a $40 billion industry, and only 1% of the world's plant species have been thoroughly studied.[16]

To a large degree, the questions pertaining to solar energy focus on which conversion technologies will contribute the most and what breakthroughs may facilitate market penetration. Photovoltaic applications are severely limited by their present low efficiencies and high production costs, though. They have the advantage of being able to use diffused light (to varying degrees). Solar Thermal Electric Conversion (STEC) processes have greater potential for large-scale application but require concentrated light. In essentially all solar applications, storage capacity presents a serious limitation to commercial applications. Since entering quantities in the reserve column means that a portion of the resource is deemed recoverable under existing technical and economic constraints, understanding issues of commerciality is essential.

WIND POWER

Wind energy also has a huge global resource base. Reserves are, again, another story. Currently, very little of humanity's electricity is wind generated. Some authors are optimistic about the potential of wind. Interestingly, few of the wind optimists refer to revisiting the largest wind machines ever built—for sailing vessels. The generation of electricity provides a much more versatile resource that is attractive to a large number of consumers. The fact remains that very little experience has been achieved in commercial generation of electricity through wind turbines. The use of this energy resource is constrained by intermittency and unreliability, invoking the need for storage, just as in the case of solar power. Wind suffers an additional constraint in that electric generation is a strong function of the velocity of the wind. Many of the windiest parts of the world are totally unpopulated, including the polar regions, the open seas, and deserts. Small-scale applications of wind power are of interest to remote consumers. Will policy decisions which transfer indirect costs of combustion energy sources to the price paid by the consumer offer the potential and incentive to develop wind applications further?

HYDROPOWER

Hydropower has come to refer almost exclusively to the generation of electricity by running water through turbines. This trend towards using hydropower for generating electricity parallels the move towards large-scale dam projects. Small-scale and mechanical (rather than electric-generating) applications have largely been left behind. However, concern for the environmental impact of large-scale dam projects may offer impetus for more small-scale applications. The total global resource base for hydropower is estimated to be approximately 95.4 EJ (90 quads) per year, with a potential to produce as much as 37 EJ (35 quads).[17] This number does not include harnessing energy from wave and tidal action, which this book includes in the hydropower category. Since water movement is a surface phenomenon, though, exploration is not an issue. Thus, only economic and technologic improvements in utilization can increase the reserve fraction of the resource base.

GEOTHERMAL POWER

The word geothermal literally means *earth heat.* Therefore, this is a ubiquitous resource, with an incredibly vast base. Reserves are constrained to relatively shallow, high temperature reservoirs, due largely to the technologic limits on preventing heat loss of produced fluids to the rocks surrounding the wellbore through which they are produced.

NUCLEAR POWER

As most authors note, a tremendous amount of uranium is distributed throughout the earth's crust. It is the breadth of distribution that limits the amount that can be considered resources. Beds of rock in which uranium concentrations are markedly higher than normal are rare. A 1% concentration of uranium in a bed makes for a very rich vein. Typically, zones with approximately .2% concentrations are commercially mined.[18]

CONSERVATION

Among any affluent people, conservation is an often underutilized alternative with great potential. Americans consume 80 times more energy per capita than many people in lower income countries, and twice as much as residents of other affluent, industrialized societies, such as Japan and Germany. With 5% of the world's population, Americans consume 25% of the world's commercial energy supplies. The startling reality is that if Americans simply brought per capita consumption into line with that of

Germany or Japan, a 12.5% savings in global commercial energy consumption would be effected. This savings represents more than twice America's total imported oil. If all of the industrialized world embarked on a committed program of energy conservation, the results could be staggering. These are of course unrealistically optimistic suggestions but serve to illustrate that the potential equivalent resource base of conservation is quite large.

A sweeping commitment on a personal and societal level would be required to utilize any large portion of the conservation resource base of the industrialized world. Americans would have to elect to purchase smaller, lighter cars and to forego some of the freedom of traveling alone in automobiles. To facilitate the latter, governments (especially America's) would have to commit to expanding mass transit immensely. Status consumption (e.g., demonstrating one's affluence by using limousines or overpowered cars, by owning ostentatiously large homes, and by indulging in excessive travel) would have to become undesirable. Our values would have to change. The changes wouldn't be easy, for who is to determine how large a home one needs or how much travel is excessive? Would government have the right to ban the most energy inefficient vehicles? For policies to succeed, they will probably need to be a reflection of a popular will to value energy more highly, whether it be for environmental concern, for national security, or for sustainability. However, such steps could replace more energy usage than all of the tankers plying the seas today, or than solar, wind, or cold fusion are likely to replace before the middle of the 21st century. These issues will be examined in detail in the final chapter of this text.

CONCLUSIONS

Table 1.3 shows how estimates of resource bases of the various energy sources, and their current levels, compare with reserve estimates. Convenience has been a determining factor in resource uti-

Table 1.3 Abundance and Availability of Energy Sources

	Production	Reserves	Resource Base (Quadrillion Btu)
Source			
Coal	93	16,200	1,060,000
Oil	133	3,538	216,000
Gas	67	3,900	25,000
Biomass	1	*	300/yr
Solar	nil	nil	4,000,000/yr**
Wind	nil	nil	8,000/yr***
Geothermal	nil	nil	1,000,000****
Hydro	21	40/yr	95/yr
Nuclear	19	2,400	10,000
Total	334		

*Biomass is a unique resource: it has a sustainable (renewable) base, but it also has standing reserves which can be and are currently being exploited.

**For solar, the gross resource base is given, which includes all solar radiation reaching the earth's surface, but note that calculations ultimately must reflect the other essential purposes that sunlight serves—warming the earth's surface, photosynthesis, etc. This author has no basis for estimating what fraction should truly be considered a resource base.

***The global wind resource is crudely calculated from Cook, p. 53, indicating that approximately .2% of total solar radiation is converted to wind by heating air masses.

****Geothermal reserves are estimated on the basis of extracting heat from an existing hot water reservoir; so these figures should be pessimistic, since geothermal has some renewal process. The figures shown, however, include lower temperature geothermal resources, which although of lower grade account for 4 times the high temperature base.

For all resources not producing substantial quantities of commercial energy, the reserves are nil (or too small to report in this format) by definition.

Sources: Edmonds and Reilly, p. 78–81, 218, 232; Hedley, p. 171; Cook, p. 53; SIPI, p. 168–170.

lization. Although development of the vast resources of solar and wind power would probably have a positive environmental impact, a serious convenience cost would be exacted because the product of these energy sources is neither readily stored nor transported great distances. Thus, if either resource is to achieve a significant market penetration, fundamental changes in consumer attitudes and preferences or dramatic changes in the relative prices will be required.

The fossil fuels are, in practical human terms, finite resources which can and are being depleted at rates greatly exceeding their renewal rates. New discoveries constantly increase the reserves and known resource bases. Although this process cannot go on indefinitely, it seems likely that fossil fuels (particularly the fluid oil and gas) will continue to dominate the world energy market through much of the next century. All current knowledge points to an irrefutable conclusion that the current consumption of petroleum products cannot and should not be sustained indefinitely. Nevertheless, all of the fossil fuels have adequate resource bases to sustain consumption levels for at least one more generation. Coal and kerogen are known to have adequate resource bases to meet current levels of demand for perhaps 20 generations, but coal is a resource left behind by industrialized societies and kerogen has never proven itself by taking a serious market share. Societies with access to technology and capital have consistently moved from resources of less energy density, convenience, efficiency, and cleanliness to resources superior in these respects for their major energy providers. (If efficiency and cleanliness must compete with energy density and convenience, the former lose. This can be seen in the move away from windmills and sailing ships. In terms of societies' dominant energy sources, though, the shifts were from firewood to coal to oil. In each of these transitions, gains were seen in all four qualities.)

Coal is certainly less convenient and less environmentally benign than oil and gas. Will the industrialized world refuse to take a regressive step, or will it pursue technology to produce improved secondary fuels from these vast primary resources? Some would say that the time has already come to move away from all combustion fuels, but to what? None of the non-combustion sources show themselves to be ready, in a practical sense, to take over for the fossil fuels, with the possible exceptions of nuclear energy and conservation.

Nuclear energy is embroiled in quasi-scientific/political/public debate as to whether it is better than any of the existing energy options. The criteria on which *better* is judged must be defined in order to consider this question intelligently.

Conservation is an alternative available to virtually every American and to most of the industrialized world. Vast amounts of energy could theoretically be replaced by conservation, but societal values appear to require change in order to accomplish this. While efficiency gains may be achieved everywhere, the affluent nations cannot expect this to effect conservation in lower income countries. Indeed, the benefits from their current paltry energy consumption need to be enhanced.

ENDNOTES

1. Jeffrey, Edward C. 1925, *Coal and Civilization*, MacMillan Co., NY, pp. 165–166.
2. Cook, Earl, *Man, Energy, Society* p. 53. Hedley, Don 1986, World Energy: *The Facts and the Future,* Euromonitor Publications, London, p. 171.
3. Ward, Barbara and Dubos, Rene 1972, *Only One Earth,* Norton & Co, p. 34.
4. Hedley, *The Facts and the Future.*
5. Edmonds, Jae and Reilly, John M. 1985, *Global Energy: Assessing the Future,* Oxford University Press, NY, p. 157.
6. ibid, pp. 76–89.
7. Hubbard, M. King 1962, *Energy Resources,* National Academy of Sciences, Washington, D.C., p. 75.
8. Giddens, Paul, *Early Days of Oil,* Princeton University Press, p. 1.
9. Yergin, Daniel 1991, *The Prize,* Simon and Schuster, NY, pp. 58–587.
10. Hedley, *The Facts and the Future,* p. 237.
11. Yergin, *The Prize,* pp. 395, 409, 480, 493, 494.
12. Abubakar, Ahmad 1989, *Africa and the Challenge of Development: Acquiescence and Dependency,* Praeger Press, NY.
13. Smil, Vaclav 1987, *Energy, Food, Environment,* Clarendon Press, Oxford, p. 26.
14. Edmonds and Reilly, *Global Energy: Accessing the Future,* p. 232.
15 Hedley, *The Facts and the Future,* p. 171.
16. World Commission on Environment and Development 1987, *Our Common Future,* Oxford University Press, NY, p. 155.
17. Edmonds and Reilly, *Global Energy: Accessing the Future,* pp. 218, 219.
18. Cook, *Man, Energy, Society,* pp. 102, 103.

CHAPTER 2

Acquisition

One common distinction between fossil fuels and renewables is how they are acquired. Exploratory technologies are generally required in order to find the energy resources trapped beneath the earth's surface (coal, oil, and gas). Kerogen (a relatively untapped fossil fuel) and uranium minerals are also extracted from the subsurface. Exploration for most renewable sources (solar, wind, and hydro), however, is a straightforward matter of observation. Yet since these resources have low energy densities, obtaining commercially viable quantities of energy depend on their capture and conversion technologies. One exception to the distinction is the exploration and production of geothermal energy which, though potentially nondepletable, is nonetheless acquired like the fossil fuels.

Combustion Fuels

COAL

Exploration • The fact that many coals contain extensive plant fossils makes clear the origin of this energy source. Since this solid fuel is often the product of paleo-swamps, exploration is focused on areas known or likely to have been swamp lands at some point in geologic time. Over these great periods of time, changes in climate, extent of the polar ice caps, tectonic uplift, and subsidence of the land itself transformed the earth's surface, sometimes leaving the fossil swamp in totally unswamp-like settings, including high in modern mountains.

Though coal is ultimately extracted from the subsurface,

exploration for coal has historically relied on direct observation. Coal seams may be seen exposed along natural cliffs and canyon walls or in road cuts. Coal is most commonly discovered in outcrops despite the likelihood that erosion may have obliterated the seam's exposed face. Geologists then look for a residual black band, referred to as *smut* or the *blossom*. Coal fragments may also be found in debris at the base of the outcropping. Exposed outcrops are most likely to be found in ravines and stream channel cuts. Natural springs in the rock face are likely to be in coal seams. Digging a few feet into the rock around the spring can uncover the coalbed itself.[1]

Once a coal deposit is identified, core-hole testing (boring) follows to map the thickness and extent of the seam and to explore for other seams. The core-hole testing is very similar to the process described later for oil and gas drilling, except that a deep coal core may be drilled to several hundred feet, as opposed to more than 10,000 feet for an oil well. Also, since the solid rock material itself is the target of coal production, rather than the fluids contained, the core recovery is less problematic. Frequently, several seams are identified at various depths in a core hole, reflecting that swamps died out and reestablished themselves at some point, as the seas transgressed and regressed.

In addition to the volume of a coalbed (or beds), the rank and quality of the coal are important, so tests or assays are made on the coal samples. Rank relates to the age and depth of burial of a coal, but a correlation between these factors and energy content is the factor of commercial interest. Quality refers to the purity of the coal, for example, how much silt and sand were buried with the decaying plant matter.

Extraction • Generally, coal is mined. It can be recovered from surface strip mines or underground mines. Each has its advantages and disadvantages, and neither is particularly safe or environmentally benign. Concern for these factors has prompted research into in situ gasification and liquefaction processes, in which the solid fuel can be converted into a fluid in the ground and recovered through a wellbore. In spite of many benefits—especially in terms of worker safety—the processes have not yet reached an efficiency level that makes them cost-competitive to contribute any significant portion of production.

Underground mining developed more than two centuries ago, as people found that coal was an efficient fuel and began to dig into seams exposed on hillsides and ravine walls. Cave-ins doubtless claimed the first fatalities in this rugged business, as men dug far enough into the seams to weaken the rock face. Then, miners used timbers to brace the tunnel-like mine shafts and dug even deeper into the mine seams. Light was soon the next problem. Miners of precious metals had encountered similar problems long before coal mining developed, but carrying torches with open flames presented new problems in the coal mines. Lower rank coals contain significant amounts of methane—which is highly explosive in the presence of air and flame. Coal mine explosions were often followed by prolonged fires from the methane which continued to seep from the small fissures and cracks in the coal. Methane provided fuel for fires that could burn for years. The advent of electric lighting diminished this problem.

Ventilation is required for a mine to reach more than a few dozen meters from the surface. Air shafts are bored from the surface into the mine, and modern mines have huge fans moving stagnant air out of and fresh air into the mine. In the past century, miners sometimes took canaries underground with them. The canaries were very susceptible to asphyxiation from methane, and a dead canary served as a warning to evacuate the mine.

Probably the most infamous problem related to underground mining is far too insidious for dying canaries to offer warning. Miners inevitably inhale coal dust, and often contract black lung disease, which actually refers to a range of disorders. The most pernicious form is technically referred to as *Pneumonocosis*. This disorder is caused by exceptionally fine silica particles—particles which are too fine to be trapped in any ordinary filter. The plant matter that formed coals was commonly deposited in waters so still that only very fine mineral matter was carried and deposited with it. This material presents a serious hazard to miners' lungs. The particles can actually be fine enough to split genes. Since the particles are so small, they are impossible to remove completely from the mine air. Although filters and dust masks cannot be totally effective, they remove some of the particles. Used in combination with improved ventilation and reasonable working shifts, the incidence of all forms of black lung disease has been reduced by more than an order of magnitude. However, it appears unlikely that this disease will be eradicated as long as miners must work underground.

Throughout the 20th century, underground mining became increasingly mechanized. At first, the move to mechanization was motivated almost entirely by increasing productivity; however, increasing public awareness of mining hazards and concern for mine safety has driven this movement in more recent years. The issue in the early days of coal mining was exacerbated by child labor. Children commonly began working in the underground mines at the age of 6 years, working 12- to 18-hour days.[2] By the last quarter of the century, a majority of mining was performed by workers operating large machinery.

Underground Mining Techniques • Coal seems are often not very thick; thus, the rooms may not be large enough for a miner to stand. The rooms are generally wide and long; they are often 10 to 40 feet wide and more than 100 feet long (Fig. 2–1a & b).[3]

In all types of underground mines, horizontal passages can be referred to as adits, tunnels, or drifts. Drifts are generally tunnels which serve to extract the coal in addition to serving as passages. Vertical passages are referred to as rises, winzes, and chutes. Rises extend upward from one working area to another, and winzes extend downward (thus, where one stands determines whether a passage is a rise or a winze). Chutes are intended as passages for coal rather than workers, allowing the product to fall from one mining area to a lower gathering and hauling area.[4]

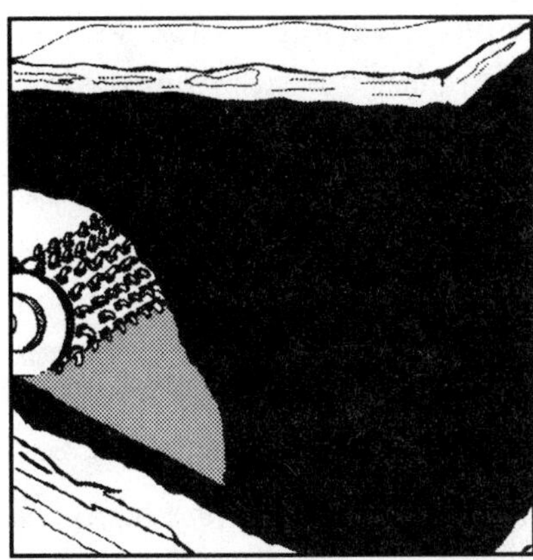

Figure 2–1a & b Methods of underground coal mining.
Figure 2–1a is a stylized representation of undercutting a coal seam. Small explosive charges placed in the overlying coal can readily cause it to collapse into the void. Figure 2–1b illustrates two rooms and an intervening pillar or stenton in room and pillar mining. Sketch by Mr. Lloyd Brown.

Room and pillar mining is one common method of underground extraction. In this case, rooms are created by the void left by excavated coal. Pillars or walls of coal, also called stentons, are left to support the roof of the mine. Roof bolts are installed between the pillars to provide integrity to the roof. The roof bolts are screwed upward into overlying beds of rock. They have sizable steel plates on the bottom which are snugged into the mine roof at about 4-foot spacing. This eliminates the need for timbers to support the mine. The supporting pillars contain significant amounts of coal—as much as 60% of the total minable ore. Once the mine excavation advances to the limits of the bed, extraction of coal from the supporting pillars begins, called *robbing the pillars.* This is done in a retreating fashion, as cave-ins follow behind the robbed pillars. Commonly, larger, more permanent tunnels and work areas are maintained, connecting the currently and previously excavated rooms and the surface. In some cases, nearly all of the coal contained in the pillars can be recovered in the robbing operation.[5]

Caving is another prevalent form of mining, which includes block caving, stoping, and top slicing. In all of these techniques, coal is undercut and allowed to collapse under its own weight, sometimes directly into a chute. In block caving, a large section has drifts mined into its base, leaving pillars behind. Then the pillars are removed (commonly by blasting), and the large block falls. The term *stoping* is a variation of step, and the mining process creates a stairstep of coal. If miners begin at the bottom corner of a coalbed and cut steps upward, it is overhand stoping; underhand stoping starts at the top and cuts steps down. In the top slicing method, a timber mat is laid to separate the ore from the capping debris. Top slicing begins at the top of a seam and works down.[6]

The longwall method of mining has two variations, known as retreating and advancing. The seam is mined in a single advancing or retreating operation, working on a long wall of coal (perhaps several hundred yards long). Waste rock, or *gob,* is used to fill and support the roof. The coal is again undercut, this time along the long face or wall, and allowed to settle. Slices may also be cut along the side to facilitate the settling. In the advancing method, the working face moves toward the boundary of the seam, with the tunnel and haulage passages being in the previously mined areas. In the retreating method, tunnels are excavated to the boundary, and the workings move back toward the main shaft.[7]

The equipment used in underground mining includes power hammers (similar to the pneumatic jack hammers seen breaking concrete on city streets), slicing and undercutting equipment, and loading equipment. A large portion of the work of an underground coal mine is breaking the coal into rubble of a moderate size that can be handled. Blasting can be an important step in mining, but normally the blast itself is not designed to break the majority of the coal. It is intended to shake loose the large blocks, pillars, walls, or slices of coal that have been cut into. The coal falls and breaks apart under its own weight and brittleness. Coal loaders, often a sort of conveyor belt with teeth or cups, scoop the coal debris up into shuttle cars where the coal is loaded and carried to the surface. The tipple (for soft coals) or breaker (for hard coals) is ideally located below the mine entrance so that coal can simply be dumped. In this facility, the coal is screened, sorted, and prepared for transport to the market.

Strip Mining • Surface mining is considerably less hazardous and less expensive than underground mining, yet its impact on the environment is much more obvious, if not clearly worse. Strip mining is similar to quarrying. The overburden of soil and overlying rock is stripped away and piled in spoils mounds. Like underground mining, the coal must be broken into rubble that can be scooped up. Blasting, drilling, and large saws perform these operations. The coal shallow enough to strip mine is relatively soft, and this can be done generally without great effort. Then the rubblized coal is scooped up, generally in a device called a dragline, and loaded into huge dump trucks.

An impressive feature of strip mining is its scale. Draglines became immense through the 1950s and 1960s, culminating in one named Big Muskie (so named because it was commissioned in Muskingum County, Ohio in 1971). The Big Muskie drags a bucket that scoops up 220 cubic yards in a single load. It has 170 electric motors, producing 40–2,000 horsepower each. Its boom reaches out 310 feet. Similarly, the trucks onto which coal is loaded carry 200 tons in a load; their tires are larger than a normal car.[8] A dragline differs from a power or steam shovel in size and the fact that the bucket is drawn by cables up and down the walking beam, rather than being hinged to its terminus. Power shovels are also busy in the strip mines, loading gathered coal into the trucks. Strip mines can effectively remove a dozen feet of overburden for each foot thickness of coal to be removed. In addition, this stripping ratio is a function of the rank and quality of the coal.

Once again, the coal is loaded into a tipple, which serves as a storage and transfer facility. The coal is typically dumped from the tipple into rail cars for transport.

At the end of a strip mine's life, the reclamation process begins. Since 1977's Surface Mining Control and Reclamation Act, law requires that the surface be restored to within 10% of original grade. This means that the slopes of the new topography must match the original slopes to within 10%. While a strip mine may tear 50 feet or more into the earth, the coal seam removed is likely only to be 4 to 5 feet thick. The spoils piles are pushed back into the hole. Graders and bulldozers work at restoring nearly original contours. Current law also requires replanting vegetation on the restored topsoil. Prudent operations call for piling soil separately from overburden rock, so that restoration can proceed effectively and tons of rock will not be mixed into the soil.

Coalbed Methane • Mitigating the dramatic hazard of explosions and mine fires in underground mining is a driving reason behind coalbed methane production. Methane gas is a chemically ideal endproduct of the anaerobic decomposition of organic material and is commonly present, to varying degrees, in coal seams. The gas is also explosive in extraordinarily small quantities in air. Thus, it is desirable to remove methane from coalbeds prior to mining. Wells are drilled with the same technology and equipment as will be described for oil and gas wells, though a deep coal may be 5,000 feet below the surface, which is still shallow in terms of modern oil and gas drilling.

A high quality coal is mostly soft, decayed biotic matter, thus there is little open pore space. What permeability exists to permit fluids to flow naturally through the coal, then, is in the natural fractures or cleats and joints. So, the coalbeds are hydraulically fractured in order to create large, permeable paths, connecting a large volume to a wellbore. The methane produced makes subsequent mining much safer and is a valuable, clean-burning product which would escape to the atmosphere even if it did not explode in the mine. Federal subsidies have encouraged the production of this unconventional gas source in the United States.

OIL AND GAS

Exploration • Exploration begins with a search for sedimentary basins which are likely to contain the massive source rocks and reservoir beds which might, in turn, contain significant accumulations of hydrocarbons. Geologists prepare basin studies, mapping the extent and ages of sedimentary rocks in a region. They consider the depositional environment or the conditions in which the sediments were deposited. Since western scientific theory holds that oil and gas are derived from decayed biota, geologists search for depositional environments where sediments were deposited rather quickly and life was abundant. Generally, attention is focused on paleo (very old) river deltas, coral reefs, and related features such as sand bars and pro-deltaic (delta front) sediments. Most sediments are deposited in aquatic environments, simply because water is the medium capable of carrying significant quantities of rock fragments or chemical salts.

In the early days of petroleum development, personal observations often turned up positive surface indications of hydrocarbon reservoirs in the form of oil seeps. The much lauded Drake well discovery was drilled near Oil Creek in Pennsylvania, which got its name from a constant sheen of oil on the waters long before Drake came along. The sheen came from a natural oil seep. Near Lander, Wyoming, a similar oil sheen on a creek led to the first Rocky Mountain discovery well being dug by hand, within 20 years of Drake's discovery. While this phenomenon was helpful in guiding early exploration, it also doubtlessly helped to instill the early, misguided notion that oil rivers flowed in channels beneath surface water rivers.

With each successive boom in the petroleum industry's inevitably repeating boom and bust cycle, explorationists walked more and more of the land, mapping regions where no scientist had gone before (at least none in their right mind). Witching for oil was popular into the early 20th century, and some of those walking with witching wands were markedly successful in identifying prospect sites. (In the upper Ohio river valley, this success may be a simple indication of the ubiquitous presence of oil and gas bearing sands. It may also be that the histrionics performed with a wand merely diverted onlookers' attention, while providing the explorationist a chance to roam the land, looking for such tell-tale indications as oil seeps. As with a stage magician, once the secret is known, anyone can do the trick.)

As exploration extended throughout the world, the easiest sites were found, and it became increasingly unlikely to make a new discovery based on such conspicuous evidence as surface oil seeps. However, you cannot conclude that regions in which hydrocarbons are not indicated by obvious, surface means are not likely to contain any significant reservoirs. Eminent oil men have erroneously condemned prospects throughout the history of the business, offering to "drink all of the oil that could be found in Texas, [Saudi Arabia, etc.]!"[9]

In the middle of the 20th century, seismic exploration became the dominant tool for identifying subsurface potential. Even if surface expressions show the presence of hydrocarbons, it is still important to obtain some idea of the possible size and depth of the reservoir from which the escaping hydrocarbons emanate. Seismic measurements can provide an image of the subsurface, working much like sonar. A noise is generated at the surface and propagated downward through layers of rock. Every time the sound wave reaches the interface between rock layers of different composition, a portion of the wave is reflected back to the surface. Microphones positioned along a line pick up a series of the sound wave reflections and record their arrivals very precisely (to the millisecond). The raw data recordings provide irregular acoustic traces for each microphone. The geophysicist working with the seismic recordings must try to match spikes representing reflections from the same bed on each microphone trace and calculate approximate depths to each bed at the point at which the acoustic wave was reflected to that microphone. The job is made more difficult by the presence of noise and multiple reflections within beds, calling for the geophysicist to spend long hours processing the data. Numerous sets of recordings are necessary for even a 2-dimensional, cross-sectional view of the subsurface. Ultimately, though, the geophysicist does compile a cross section, and determines which beds are most likely to be reservoir-type rocks and which are more likely to be shales.

Sound waves travel fastest through hard, dense material. If a sandstone is well-consolidated (that is, it has good grain-to-grain contacts even if it has substantial porosity), the acoustic wave is provided a good, fast route. Shales tend to be composed largely of clay platelets laying on one another, with varying amounts of water hydrating each platelet. Thus, shales commonly have slower transit times than sandstones, which are in turn not as fast as limestones. A poorly consolidated sandstone complicates the picture. As the sound wave passes through nonrigid grain-to-grain contacts, it attenuates, or loses time and energy. Nevertheless, the geophysicist, commonly working closely with a geologist, can present a sufficiently detailed and accurate picture of the subsurface to convince senior executives of major oil and gas corporations to invest perhaps $10 million for a single international exploratory well.

Satellite imagery can be utilized in the preliminary stages of modern exploration to indicate the likely presence of subsurface hydrocarbons. Surface topography reflecting subsurface features can be seen, and definitive indications of hydrocarbons sometimes may be observed. Satellite infrared images can identify *hot spots* where escaping methane is present in the atmosphere, because methane is a strong absorber of infrared radiation. Unfortunately, the identification of hot spots has not proven to be a successful exploration technique.

An exploratory well is known as a wildcat, which is a very high risk venture. A long-standing rule-of-thumb was that about 1 wildcat in 17 was a commercially successful discovery well. It is true that advances in technology have led to improvements in the success ratios, as has the increasing knowledge of geologic basins in explored regions; however, *rank wildcats* in new geologic provinces still remain extraordinarily high risk ventures.

The Meaning of a Commercially Successful Discovery Well • Wells which fall into this category identify sufficient producible hydrocarbons to justify all of the development expenses required to tap the discovery, produce it, and deliver it to market. Thus, a single well which produces 100 MCF/day of gas from a depth of 3,000 feet, near an existing pipeline in the United States, could reasonably be a commercial discovery. On the other hand, a wildcat may be unsuccessful even if it tests an initial production of 10,000 MCF/day, and is followed by several confirmation wells, which identify 1 TCF of gas in place. Perhaps the production is from too deep or a difficult drilling horizon for the field to be *drilled-up* cost effectively. Perhaps the discovery is in deep or hostile waters (e.g., the Arctic Ocean). Or perhaps the discovery is too remote to adequate markets. One very large Third World field was discovered, but the companies estimated that the pipeline required to transport the production across the mountains to an adequate market would cost more than a billion dollars (U.S.), which made companies write off even that very large discovery as sub-commercial.

Drilling • Successful exploration requires drilling the first well. The process is essentially the same as for drilling a development well in a known field—or, for that matter, drilling a well to tap geothermal energy or conducting in situ leaching of a uranium deposit.

Early technologies of drilling can be classified as *percussion drilling.* More than 1,000 years ago in China, people drilled with spring poles. In this form of drilling, a lever arm is fixed across a fulcrum several feet off the ground. (A sapling can be used for the lever arm and a larger tree for the fulcrum, as in Figure 2.2.) A chisel-like bit is tied to one end of the lever arm, and a stirrup is attached to the other. A person kicking down on the stirrup can work the lever arm, imparting a sharp, vertical reciprocating motion to the bit. Wells have been drilled to more than 1,000 feet with this method.[10]

This technique was extended in the 1800s by using an engine and a crank arm to raise and lower the end of the lever arm, and a derrick was used to run a long, continuous rope or cable over an elevated pulley. This method was developed by salt drillers. In 1859, Edwin Drake hired a salt driller known as Uncle Billy to drill an oil well near Titusville, Pennsylvania. Uncle Billy's driller's log indicated that the "bit fell into a crevice from 62 to 67

Figure 2–2 Kicking down a well using a spring pole.
Stepping into the stirrup-type ropes at the well allows the tools suspended from the rope to fall, striking the bottom of the hole. The resilience of the sapling pulls the tools back up. Sketch appearing in *McClure's Magazine,* February, 1903, from Paul Giddens, *Early Days of Oil.*

feet."[11] Oil and salt water began to flow from the well. Thus was born the first American oil boom. Essentially the same technology was used for 60 years, and is still used in relatively shallow applications today. During the 1920s, some drillers experimented with rotary drilling systems in which a bit was screwed onto the end of a joint of pipe, which in turn was rotated at the surface, and then the pipe conveyed the rotation to the bit. When the hole was as deep as the length of pipe would allow, another length of pipe was screwed onto the upper end of the pipe before drilling resumed. Rotary drilling received tremendous stimulus from the advent of the *tri-cone roller bit* introduced in 1924 by Howard Hughes. The solid bits had been impractical because they would wear rapidly and had a tendency to cause the pipe to fail in torque (twist off). The new roller-cone bits, still in use today, turn more freely, continually alternating cutting teeth to the rock surface. They also have holes in the middle which enable fluid to be circulated down the drillpipe, and out the holes in the bit, to jet rock cuttings away from the teeth and carry them by the circulating fluid back up the annulus between the drillpipe and the hole to the surface.This became the dominant drilling mode by 1940.

By the 1970s, rotary drilling technology permitted reaching more than 30,000 feet, nearly 6 miles, into the earth's crust. (The thickness of the earth's crust varies from around 100,000 to 160,000 feet; it is clear that drilling achievements remain significantly short of penetrating the crust to the mantle.) At such depths, the fluids found in rock pores can be under tremendous pressure.

Drilling Rig • The rig itself consists of the derrick, a prime mover, the surface mud system, pipe storage racks, a hoisting system, rotating equipment, the blow-out prevention system (BOPs), and a *doghouse* for the crew to step out of the weather and to house some instrumentation (Fig. 2-3).

The derrick is the tower-like structure most readily associated with drilling. It provides a hoisting structure and a place to store drillpipe during trips. Each joint of pipe is nominally 30 feet long, but to expedite drilling operations, modern drilling rigs normally use trebles derricks, which are tall enough to stand 3-joint (90-foot) sections of pipe in the derrick. Shorter singles and doubles derricks are not preferred because the operation of pulling pipe out of the hole for such purposes as changing a worn-out bit is much more labor intensive.

The drilling fluid is called mud, and it has three main functions. It lubricates the bit, it lifts cutting out of the hole, and it maintains necessary pressure balances to control the well. The drillpipe connects the bit to the surface, transmits torque to the bit, and conveys drilling mud down its interior and back up its exterior (the annulus between pipe and hole).[12]

With modern technologies, drilling as much as 5,000 feet of hole on a single bit run is possible, but the bit wearing out in a long stretch of drilling remains likely. If penetration rates cause the driller to suspect that the bit is wearing out, he initiates a trip to replace it. Pulling the pipe out of the hole is called *tripping out*,

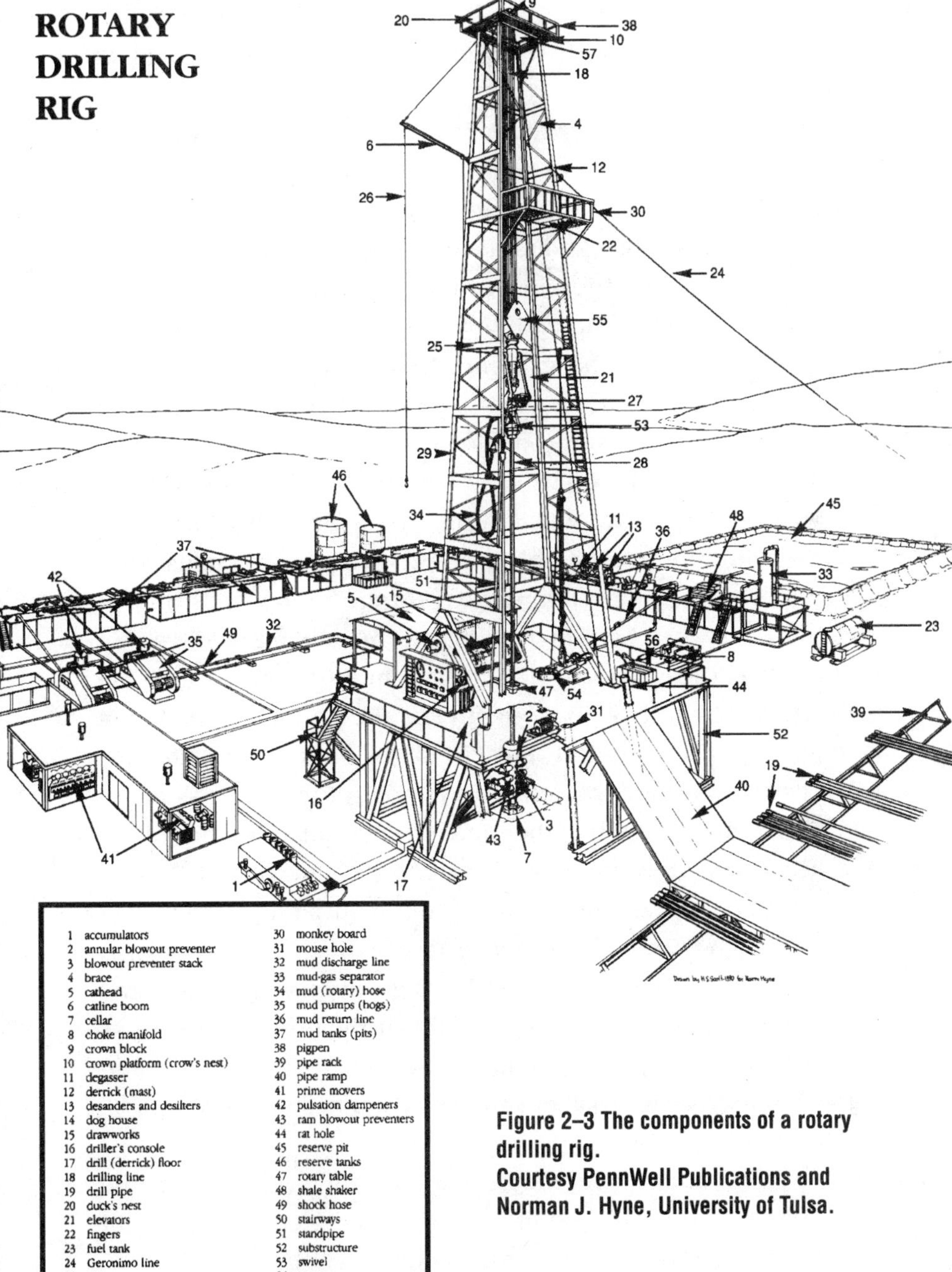

1 accumulators
2 annular blowout preventer
3 blowout preventer stack
4 brace
5 cathead
6 catline boom
7 cellar
8 choke manifold
9 crown block
10 crown platform (crow's nest)
11 degasser
12 derrick (mast)
13 desanders and desilters
14 dog house
15 drawworks
16 driller's console
17 drill (derrick) floor
18 drilling line
19 drill pipe
20 duck's nest
21 elevators
22 fingers
23 fuel tank
24 Geronimo line
25 girt
26 hoisting line
27 hook
28 kelly
29 leg
30 monkey board
31 mouse hole
32 mud discharge line
33 mud-gas separator
34 mud (rotary) hose
35 mud pumps (hogs)
36 mud return line
37 mud tanks (pits)
38 pigpen
39 pipe rack
40 pipe ramp
41 prime movers
42 pulsation dampeners
43 ram blowout preventers
44 rat hole
45 reserve pit
46 reserve tanks
47 rotary table
48 shale shaker
49 shock hose
50 stairways
51 standpipe
52 substructure
53 swivel
54 tongs
55 traveling block
56 trip tank
57 water table

Figure 2–3 The components of a rotary drilling rig.
Courtesy PennWell Publications and Norman J. Hyne, University of Tulsa.

going back into the hole is *tripping in,* and the whole process is called a *round trip*.

During a trip, the derrick man stands on the monkey board (the small platform about two-thirds of the way up the derrick) to handle the top ends of the pipe stands. The hoisting apparatus includes the crown blocks and traveling blocks. The crown blocks are so named because they are set at the very top of the rig, while the traveling set move up and down as they raise or lower pipe. While drilling, the kelly hangs from a hook on the traveling blocks. The kelly is a square-sided joint of pipe, which is always at the top of the drillstring. It fits into a square hole in the kelly bushing, which is affixed in the rotary table. The rotary table can be imagined as a huge sprocket wheel turned by a chain drive from the engines. When the rotary table turns, the square kelly must turn with the kelly bushing—this imparts torque to the drillstring, of which the bit is the bottom-most piece. Thus the bit turns so that it can chew through rock. The drilling mud is pumped into the drillpipe through the kelly hose; it exits the bottom of the string through nozzles in the bit and circulates back up the annulus between the pipe and the walls of the hole. (The bit at this depth may be 12.25 inches, while the pipe is 5 inches.) At the top of the hole, the drilling mud flows out through a connection at the top of the surface casing that leads it to the mud pit. The return line dumps the mud onto a shale shaker (a vibrating sieve-like device) to remove rock cuttings before allowing the mud to reenter the pit, where mud pumps send it back down the hole. The mud pits are large enough to give the mud a reasonable residence time, during which fine particles fall out and any entrained gas escapes.

Drilling Procedure • As the bit chews through more rock, the driller eases off the brake on the traveling block, letting the weight of the drillstring pull the bit farther down, trying to keep a relatively constant weight on the bit. Once the top of the kelly reaches the kelly bushing, the driller raises the drillstring until the connection between the bottom of the kelly and the top-most joint of pipe is exposed. The roughnecks (drilling crew) set slips in the kelly bushing, which have teeth like those of a pipe wrench pointed up. The four sides of the slips fit into the square kelly bushing and are tapered so that as the weight of the pipe pulls them down; they are also pulled more tightly against the pipe. Thus the pipe is held firmly, the kelly is unscrewed, a new joint of drillpipe is picked up and screwed onto the top of the string, the slips are removed, the pipe lowered to its new top, the slips reset, and the kelly reattached. Drilling proceeds again.

Prudent operations, if not law, require setting casing (relatively large-diameter pipe) to a depth to protect potable aquifers from contamination and to ensure that weak formations are not exposed to dense muds. Before starting to drill the well, the engineer may have estimated that it would be necessary to set one intermediate string of casing before reaching the 12,000 feet projected ultimate depth, to protect formations between the bottom (shoe) of the surface casing and drilling depth from dense mud. Most commonly, it is desirable to drill to total depth (TD) with a 7.875-inch bit. To do this, the intermediate casing above it must be larger—perhaps 10-inch casing. In order to set 10-inch casing, the hole must about 14 inches to allow room for an annular cement sheath. The surface casing must be about 16 inches to permit the 14-inch bit to pass through it, meaning, in turn, that the first bit would need to be about 20 inches.

If it is determined that intermediate casing is to be set at 10,000 feet, the bit is tripped out at that depth and 10,000 feet of the 10-inch casing are run into the hole, with a *guide shoe* on the bottom joint, which is heavy, tapered pipe, to prevent gouging into the side of the hole. It also has holes above the bottom to permit fluids to circulate even when it has landed on the bottom of the hole. Perhaps one joint above the guide shoe is a float collar, which has a ring inside it, somewhat smaller diameter than the casing ID (internal diameter.) When the casing has been landed, cement is pumped into it like mud is pumped into the drillpipe. The engineer has calculated the amount of cement required to fill the casing-hole annulus and some reasonable excess factor to ensure that the annulus is adequately filled. With the last bit of cement, a rubber wiper plug is released, which in turn is followed by water. The water forces the cement and plug ahead of it. Like the mud during drilling, the cement circulates down inside the pipe and back up around the outside. When the plug hits the ring, it seats, causing the pump pressure to rise dramatically. At this point, the inside of the casing is full of water (except the bottom part between the float collar and the shoe), and the annulus is full of cement. After intermediate casing is set, a smaller bit is used to drill out the bottom of the casing.

Drilling Problems • Unfortunately, there are so many variables in the physics and mechanics of drilling that wells are not often drilled without problems. One of the worst problems is when formation fluids escape violently from the well—a blowout. In overpressured conditions, the formation fluids are under greater pressure than would be exerted by a normal hydrostatic head. Thus, if such a formation is drilled into with plain water drilling mud, fluids flow from the formation into the lower-pressured wellbore. This displaces drilling mud. If the entering formation fluid is of lower density than the mud (e.g., gas), this displacement decreases the pressure in the wellbore, permitting a greater influx of formation fluids. Meanwhile, the gas bubbles up through the drilling mud, expanding, and displacing ever more mud. This snowballing condition can be disastrous if the gas reaches the surface and is able to blow out around the drillpipe. The condition in which underbalanced conditions permit formation fluid entry into the wellbore is called a kick. If those formation fluids reach the surface unchecked, it is called a blowout. A gas blowout has a high risk of igniting in spectacular fashion. Blowouts reached the popular press in 1991 when retreating Iraqi soldiers intentionally opened many producing Kuwaiti oil wells to the atmosphere and ignited them. Teams of blowout control and well-fire specialists from around the world converged on the crisis, coordinated by the most famous of them, Texan Paul 'Red' Adair.

Tripping out of the hole is one of the highest risk times for experiencing a kick. Since the bit is larger than the pipe, it acts like the plunger on a pump and tends to swab mud out of the hole, reducing the hydrostatic head of the overall mud column, and creating short-term, very low pressures below the bit as the bit lifts the mud column above. If an unsuspected overpressure is drilled into, the well may have begun to kick already. When swabbing increases the formation flow into the well, it can be a very dangerous situation. As mentioned earlier, gas entering the wellbore displaces more mud and lightens the column as it moves towards the lower pressures of the surface, aggravating the problem. On the trip in, the bit pushes mud ahead of it, causing surge pressures in the hole which could fracture low-pressured formations.

The driller probably first becomes aware of the kick when the mud pit alarm sounds. The alarm is very much like the float on a toilet tank. If more fluid enters the mud pit, it raises the float to an electric contact, triggering an alarm sufficient to rouse the dead. The driller can adjust the sensitivity of the alarm. As sometimes happens on drillships in rough seas, the driller may turn the alarm off altogether. If for such human error, or shear bad luck, the initial influx of formation fluids goes undetected and a large volume of gas is swabbed in during the trip, the kick can be very serious. The driller likely sends one of the roughnecks to look at the pit when the alarm sounds. (Another practical problem is that during a trip, all of the crew is busy, and the driller may not dispatch someone immediately.) In the worst of conditions, the mud has been exposed to gas invasion for some time and the crew member will report that the pit is frothing with gas like a freshly drafted beer.

At this point, the drilling crew becomes engaged in active blowout prevention. The engineer opens the choke on a separate blowout mud return line, the driller picks the bit up off the bottom of the hole, and tries to position the body of a joint of drillpipe in the BOPs and close the pipe rams. The pipe rams are two halves of a hard rubber doughnut forced against the pipe to seal the annulus. The choke is closed slowly, sealing the annulus side of the well while the mud pumps control the inside of the pipe. The engineer can determine the mud weight required to kill the kick by reading the pressure on the standpipe (casing). By converting the pressure to an equivalent mud weight at bottomhole conditions, the amount of extra weighting required can be determined. Additional weighting agent (e.g., barite) is added to the mud pits and pumped into the well to make the new mud denser. In the case of gas entry, the process is complicated by the tendency of the gas to expand as it rises and further decrease the annular pressure. An experienced engineer and crew are likely to be able to accomplish it, if nothing else goes wrong.

What can go wrong? The pipe rams might fail to seal. That is why the BOP stack on deep wells almost always has 100% redundancy of the basic pipe rams. Sometimes the drillpipe gets stuck, and the driller is unable to raise or lower it to get the body of the drillpipe positioned properly in the BOPs. Then the Hydril is employed. It inflates with hydraulic fluid to seal around whatever shaped object is present (larger diameter pipe collar, square kelly, piece of junk that fell in). If that fails, there are blind or shearing rams. The massive hardened steel blades driven by hydraulic force on modern rigs can shear through anything in their path, sealing the well but also preventing the circulation of kill mud.

Perhaps the worst additional problem is to lose circulation while fighting a kick. Since shallower zones are likely to be at a lower pressure gradient than the deeper zones, the pressure transmitted up-hole by the kick can force mud into those low pressure zones. Since the lowest pressure exists at the shallowest point of the open hole, which corresponds to where the shoe of the deepest casing exists, this condition is commonly referred to as *breaking down the shoe*. It can also be called *lost circulation,* since some of the circulating fluids are lost or *an internal blowout* for rather obvious reasons. Trying to circulate kill mud with lost circulation is like rowing a boat with one oar out of the water. The fluid loss in the wellbore must be controlled before the kick can be stopped. Lost circulation materials are added to the kill mud and circulated to plug the weak zone. They might be such exotic materials as walnut hulls, cellophane, etc. If circulation is not restored, the gas may enter the lost circulation zone and then fracture and/or percolate up through the overlying sediments—blowing out away from the well. This most probably means that the rig and the existing hole are lost, if not the crew. Thus, the race to prevent an imminent blowout is urgent, but also meticulous and deliberate.

Fortunately, blowouts are rare. Of 19,485 wells drilled in the United States from 1971 to 1989, only 82 drilling wells blew out. (Another 44 wells blew out during other production or completion operations.)[13] As the discussion of pressures demonstrated, the risk of blowout increases substantially with drilling depth.

Other less dramatic problems include losing drillpipe or other equipment in the wellbore. This is not uncommon because the monstrous forces employed to turn as much as six miles of steel pipe can break the pipe. Additionally, the drilling mud being circulated up the annulus tends to erode or wash-out soft rock formations. Large wash-outs can be difficult to cement properly and can serve as small caverns in which downhole equipment can become lodged. These and many other problems can be resolved with only a loss of time and expenditure of money.

Post-Drilling Evaluations • Unfortunately, only after a wildcat is drilled (and often followed by several confirmation wells) is it possible to know what the subsurface contains. Because it is such a risky venture, it is common to *plug and abandon* (P&A) a wildcat regardless of what it finds. The cost of having materials on location to complete an exploratory well as a producer is so high, relative to the likelihood of that being the appropriate decision, that it is normally prohibitive. Some evaluation is conducted continuously as the well is drilled. The more costly the total exploration program, the more comprehensive the evaluation is likely to be at every step. Routinely, a geolograph records basic drilling parameters such as rate of penetration, weight on the bit, and rotations per minute from which relative hardness and pressures of formations penetrated can be deduced. A geologist may be on site, observing rock cuttings for signs of oil stain and perhaps microfossils to indicate the age and depositional environment of the rocks. Many times a mud-logging unit is employed to sample and analyze the mud for the presence of dissolved hydrocarbons.

Since the drilling mud normally invades permeable formations and flushes formation fluids away from the wellbore, relatively little is known of the in situ fluids, even upon completion of drilling. Sometimes, zones of interest are drilled with a special doughnut-shaped bit, cutting a cylinder of rock which can be recovered—rather than crushing the formation between the teeth of roller cones. This process is called coring. The cylindrical core enters special drillpipe, called a core barrel, as the bit cuts deeper. When the pipe and bit are tripped out, the core is recovered (hopefully) in the core barrel. Normally, the core is around 3 to 4 inches in diameter, and sometimes several hundred feet of formation are cored. The core provides a physical sample of the formation, which can be measured and examined extensively on the surface. The sample, however, has most likely been altered significantly by mechanical jarring, the mud, and exposure to lower pressures and the atmosphere. One extraordinary measure to extract the core in an unaltered (virgin) state is lining the core barrel with a pressurized rubber sleeve. Even if this process is successful, one expert in formation evaluation observes that what you have is an excellent sample of the one part of the formation which is no longer there. This statement is not entirely facetious because the formations are often not homogeneous, and heterogeneities may not be accurately reflected in the small core sample. Nevertheless, cores can be important. The geologist can observe patterns in the rock that reveal the formation's depositional environment. Porosity can be measured rather directly. Fluid flow experiments can be conducted to determine the permeability, as well as many other tests. Of course, all of these tests take time—time during which a drilling rig may be sitting, waiting for a decision of whether to complete the well as a producer or to P&A it as a dryhole.

The immediate decision is normally based on well log analysis. Once the drilling has reached TD, the pipe and bit are tripped out, and logging tools (sondes) are lowered into the hole on heavy, electrical cable. The sondes measure three types of formation properties: nuclear, acoustic, and electrical. From these measurements, estimates can be made of the properties of interest, which are lithology, porosity, and saturation. Lithology refers to the rock type and is a preliminary, crude indicator of permeability. Potentially permeable reservoir rocks are sandstones, limestones, and dolomites—not shales. The two primary measurements of lithology are the natural gamma ray log and the SP log. (SP can stand for spontaneous potential or self potential, depending on the author.) The gamma ray log is rather like a Geiger counter; it measures and records gamma radiation naturally emitted by the formation. The naturally occurring isotopes likely to be found that contribute gamma radiation are potassium-40, thorium, and two daughter products of uranium. Potassium is the most common of these and is a constituent of most clays, and clays are primary constituents of shales. Sands are composed of silicon dioxide, limestones of calcium carbonate, and dolomite of calcium and magnesium carbonates. Thus, it is normally assumed that radiation is proportional to potassium count, which is proportional to clayiness or shaliness. The log analyst looks for zones with little natural radioactivity. The SP log works on the principal that drilling mud invades permeable formations. If the mud is of different salinity than the formation waters, ions tend to migrate to establish an osmotic balance. In *normal* conditions, formation waters are more saline than the mud. Thus, impermeable formations have more salinity in their original fluids than in the mud, and salt ions tend to move towards the boundaries of the formations. Since cations are smaller than anions, this creates an electrolytic imbalance across the formations (like a battery). The voltage level indicates rather qualitative differences between permeable reservoir rocks and impermeable shales.

Porosity • The next step is to determine porosity in the zones identified as potential reservoir-type rocks. The three types of measurements used are formation density, neutron, and acoustic velocity. The formation density sonde (tool) fires gamma rays into the formation. As gamma rays interact with the electron cloud, they give up some of their energy to excite the electrons. The electrons soon lose their vicarious excitement. The returning radiation is a function of how many electrons the outbound radiation found to interact with. The number of electrons per unit volume is a function of density. The more dense a formation is, the less porous it is. In order to calculate porosity accurately from the density log, the rock grain and fluid densities must be assessed with reasonable accuracy. Sandstones have grain densities of about 2.65 g/cc (grams per cubic centimeter), while limestone densities average 2.71 g/cc and dolomites 2.87. Water density is normally between 1 and 1.1 g/cc, with oil densities generally close behind in the range of .8g/cc. Calculations are normally based on a fluid density for water. If oil is present, the difference will be modest, but gas is so much less dense that its presence will cause the density calculations to overestimate porosity considerably.

The neutron log fires neutrons into the formation. Neutrons interact with formation materials, but they tend to be captured by entities of essentially equal mass, which would be hydrogen nuclei. The formation fluids contain hydrogen, while the rock materials do not. Thus, the concentration of hydrogen is proportional to the porosity. However, in the case of natural gas—since the gas is of low density—the hydrogen-bearing gas molecules are farther apart than in liquids, causing the neutron log to underestimate porosity in gas zones. This fact can be used in combination with the density log's tendency to overestimate porosity to identify the presence of gas.

The acoustic log operates on the principle that sound travels faster through dense solids than through fluids. The time it takes a sound wave to pass through the formation to a microphone is proportional to the porosity. While geophysicists are fond of the acoustic log because its measurement relates strongly to the seismic measurements made from the surface, many borehole petrophysical specialists prefer the density and neutron logs for the precision of their responses to varying mineralogy. All of the porosity logs are strongly affected by variations in the formation mineralogy and can require very sophisticated approaches to identify the mineral contents accurately and make appropriate corrections to the calculated porosity.

Saturations • Once porosity has been estimated, the next step is to determine saturations. Since oil and gas virtually always occur with some amount of water, it is important to evaluate the water saturation. Formation waters almost always have some dissolved salts, making the water electrically conductive. Generally, neither the rocks nor the oil and gas are conductive, so the electrical conductivity is proportional to the water saturation. In 1948, G.E. Archie developed the seminal equation from empirical observations, which is commonly used in modified form today. This equation relates the water saturation in a formation to its porosity and electrical conductivity, with terms representing the shape of the pore spaces and the water film on the grains, as shown in the Appendix. This equation makes intuitive sense: The measured resistivity of a porous formation is proportional to the resistivity of the formation water, and inversely proportional to its overall volume in the region of the formation being measured (porosity times saturation in the pore space yields bulk volume fraction). Therefore, the saturation of the conductive fluid (water) is proportional to the water's resistivity, and inversely proportional to the measured resistivity, and the pore space. The pore space not occupied by water is assumed to be occupied by oil and gas.

The measurements are far from perfect, being made through a relatively slim hole, perhaps two miles or more in length. The tools are constrained by bed resolution (how thick a piece of the formation they read) and by diameter of investigation (how far away from the wellbore they can read). The calculations are constrained by a large number of assumptions, because the measurements are by nature underdefined, which is to say that many more variables are present in real rock formations than can possibly be

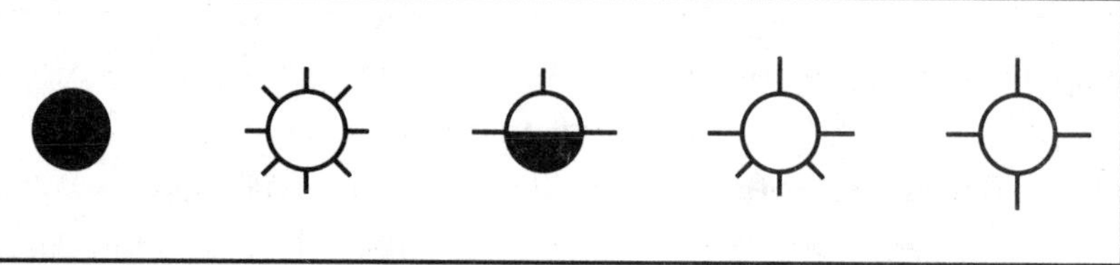

Figure 2–4 Mapping well symbols.
Oil Well, Gas Well, Dry Hole with Oil Show, Dry Hole with Gas Show, Dry Hole

measured. The calculations can be no better than the assumptions upon which they are based. Partially in recognition of this fact, one seasoned geologist has remarked, "The best place to look for oil is in an oil field." In spite of the problems associated with the measurements, a decision of whether to complete a well as a producer or to plug-and-abandon it is normally based on log analyses. Calculations are commonly done on site while the rig stands idle.

If the well is not deemed to have sufficient amounts of oil and/or gas present to be produced profitably, the well is plugged and abandoned and marked as a dry hole. The geologist places a dry-hole marker (an open circle with four rays extending towards the four points of the compass) on the map. Technically, if the well discovers sub-commercial quantities of hydrocarbons, the geologist should mark the spot on the map with oil or gas *shows.* (The different map markings are illustrated on Figure 2.4.) It is not uncommon for maps simply to have the basic dry-hole indication, as the geologist must exercise judgment as to whether oil and gas shows are significant enough to merit indication on the map. This judgment can (and some would certainly argue should) be influenced by economic factors. A well in the Central African Republic, 1,000 miles from ports or major markets, is likely to require a much larger show to be so identified on the map than a similar unsuccessful well outside of Tulsa, Oklahoma. The point here is that the context has a strong bearing on the results. Reports of exploration results are not absolute.

Completions • A well can be plugged and abandoned (as wildcats and dry holes conventionally are), or it can be completed either as a producer or an injector. If the well is plugged and abandoned, bridge plugs are set at intervals to seal the well. The bridge plug, like a packer, is a device lowered into the well which expands to press rubber seals against the wellbore or casing walls. Most often, the packer is mechanically operated, with opposing angled sides pushing the two sections of the packer apart when the pipe is twisted. The bridge plug provides a solid seal across the well, while a packer is part of the tubing or pipe which seals the annulus but permits flow inside. Cement is poured on top of the plugs. The intervals are determined to prevent formation fluids from moving up the wellbore to contaminate shallower zones, especially potable aquifers, as well as preventing fluids from reaching the surface and causing the abandoned well to blowout. In some cases, the entire well is filled with cement. A marker is placed on the surface to indicate the well's location, which is also marked on the geologic maps.

When a well is completed as either a producer or injector, casing is run to the bottom and cemented in place. Sometimes, the casing is set just before drilling into the top of the productive formation in an *open-hole* or *barefoot completion.* This may be done if the formation is likely to be water sensitive or otherwise significantly damaged by the cement. Another variation is to set a slotted liner. The slotted liner is casing with holes or slots precut to permit fluid entry. Conditions for setting a slotted liner include: unconsolidated (loose) formations in which a gravel pack (really sand) is placed around the liner to restrain formation particles carried with the production fluid or to ensure unrestricted flow from an interval that may be thinly laminated with sandstone and shales. In the conventional completion, casing of approximately 2 inches smaller than the hole diameter is lowered in and set on the bottom of the hole at TD and cemented in place. If the casing extends all of the way to the surface, it is called casing; if casing doesn't reach the surface, it is referred to as a liner. In deep wells, where several strings are run, it would be wasteful to have perhaps 5 concentric strings of pipe cemented at the top of the hole.

In such a case, the bottommost, or production string, is likely to be set as a liner, extending only several hundred feet up into the bottom of the next deepest casing. It is always important to have significant overlap between casing strings to ensure a sound hydraulic seal and prevent fluid movement around a casing shoe.

Once casing is successfully cemented in place, the hole is entirely sealed, such that no fluid can enter or exit. The whole purpose of drilling the hole, though, was to produce the oil and gas from the formation, or inject fluids into a producing reservoir. So, at this point, the well must be perforated to establish communication with the reservoir, requiring that holes be made through the pipe and cement sheath. At depths shallower than 4,000 feet, the holes are normally made by guns literally firing bullets through the pipe and cement. The perforating gun is a cylinder with a series of chambers spaced around it. The bullets are loaded in the chambers and fired electronically when the gun is positioned across the desired zone. The spacing of the chambers is phased so that the shots open holes all of the way around the casing, and that the shots approximately balance each other's recoil. Knowing when the gun is correctly positioned requires some perforating depth control, most commonly afforded by a gamma ray logging tool incorporated in the gun assembly, which can be correlated to the open-hole gamma ray log. (Gamma rays are only slightly attenuated by the steel and cement.) If the reservoir is deeper than about 4,000 feet, bullets are generally not very effective because there is so much confining pressure in the cement and rock through which the bullet is expected to force its way. In these cases, shaped charge, or jet, perforating is the norm. Here, a plastic explosive is packed into each chamber in a shape that causes the pattern of its burn to be a relatively narrow jet straight out into the formation. It works in much the same manner as a bazooka shell burns through the wall of a tank. Whether bullets or shaped charges are used, several shots are normally fired at each foot of the productive interval. Four shots per foot was something of a standard for several years. Testing in the 1980s suggested that significant fractions of shots are ineffective, causing some companies to increase their routine shots per foot. The need for numerous shots is especially seen in thinly laminated, sand-shale reservoirs, where 12 shots per foot is common.

Stimulation • The well's initial production rate may not be satisfactorily high. Since a remarkably large share of oil and gas development expenses are incurred at the front end of a project, it is especially important for the well's early production rate to be great enough for the early proceeds to repay the initial investment. A well may, however, produce at a rate that is adequate to meet economic considerations, but at a rate well below its apparent potential, due to formation damage. Since the wells are commonly drilled overbalanced, drilling mud invades permeable formations—such as the productive reservoir—often leaving mud solids in the pore space, which obstructs permeability. Even if the mud does not damage the reservoir, cement from the production casing is likely to block pores near the wellbore. In any of these circumstances, it is desirable to stimulate the well to increase or restore the reservoir's natural permeability.

Fracturing, inducing cracks in the rock, is a prevalent stimulation technique and probably the oldest. The earliest efforts involved detonating as much as several gallons of nitroglycerine in the well—called shooting the well. This ideally created a small cavern in the reservoir rock around the wellbore, and was properly done in an open (uncased) section of hole. Controlling the shot—even getting it into the well—was difficult and dangerous. There was a popular saying in the oil patch in the days of shooting. "I've known a lot of shooter's widows, but never a shooter." When successfully accomplished, though, the explosive fracturing job did increase the near-well permeability and production rate considerably.

While shooting is still performed occasionally, this practice has been largely replaced by hydraulic fracturing. The hydraulic fracturing job consists of pumping viscous fluid into the wellbore under high pressure. By using a very viscous fluid, the reservoir's ability to take the fluid in is decreased, and it fractures more readily.

Here is an intentionally induced variation of the phenomenon described earlier as *breaking down the shoe*. In order to keep the fractures open as high permeability conduits for produced fluids, proppants are normally a part of the fracturing job. The proppant may be sand particles, small ceramic beads, or aluminum particles. Fracture design and proppant selection are complicated by the enormous pressures which try to close the fractures—breaking or imbedding many proppants in the formation.

The other most common form of stimulation is to acidize the formation. Generally, hydrochloric acid, or some combination of hydrochloric and hydrofluoric acid, is pumped into the formation to dissolve materials that may be blocking the flow of fluids. Hydrochloric acid alone dissolves natural carbonate cementation between sand grains, as well as contamination near the wellbore from cementing. It also dissolves a fraction of a limestone formation, enlarging the permeable channels near the wellbore. The addition of hydrofluoric acid also dissolves mud particles and more resistant carbonates. The most common application of acidizing is in carbonate reservoirs or sandstones with natural carbonate cementation.

Sometimes hot fluids are pumped into the well to dissolve tars and parrafins. Emulsifying agents may be pumped in to break up a *water block*—the case where the water saturation near the wellbore approaches 100%, blocking hydrocarbons from flowing.

Unfortunately, stimulations frequently fail. An induced fracture may extend too far down, establishing communication with a large aquifer. Acid may react with a secondary material, producing a flocculating precipitate. Displacement water can create a water block. Many wells have been ruined through these and other damage mechanisms from failed stimulation attempts. There are far more variables in the composition of the reservoir and its production mechanisms than can possibly be measured. An attempted stimulation can do more harm than good.

Production • Initially, most wells do not need to be pumped; they begin producing as flowing wells, and almost all gas wells flow throughout their productive lives. The pressure at which the formation fluids are found is typically at least equivalent to the hydrostatic gradient of saltwater (normally pressured formations). Since oil is lighter than water, a column of oil filling the well exerts less pressure than exists in the formation, so it flows. Gas being even much lighter than oil flows readily, and the gas that is commonly dissolved in oil helps the oil wells to flow at higher rates. When the pressure in the formation is no longer great enough for an oil well to flow, artificial lift techniques are normally employed.

The fluids flow through the tortuous (crooked) pore paths between grains toward the wellbore's low pressure zone created by withdrawing fluids. As more fluid is withdrawn, the reservoir pressure near the wellbore declines almost immediately, and a pressure reduction wave spreads slowly through the reservoir. As the reservoir pressure declines, so does the production rate. When production declines to a certain point, artificial lift is considered for an oil well. By imparting energy to lift the fluid weight in the wellbore, wellbore pressure can be decreased, thus increasing the pressure difference from the formation to the wellbore, and the flow rate. (The equations are developed in the Appendix.) Gas is so light that its weight in the wellbore is negligible, so artificial lift is seldom appropriate.

The most common artificial lift method is sucker rod pumping. In this method, a plunger pump is placed at the bottom of the tubing and connected to the surface by a string of rods. The rods were once made of hickory, with steel threaded couplings riveted on each end to permit screwing the rods together. By the early part of the 20th century, all-steel rods replaced the hickory bodied ones, and at the writing of this text, fiberglass-bodied rods are gaining popularity. A lever system on the surface raises and lowers the rods, in turn lifting and dropping the plunger at the bottom. Each time the plunger is lifted, the fluid above it is lifted towards the surface. Momentarily, then, the pressure below the plunger drops dramatically, allowing more formation fluid to enter the wellbore. The lever system is typically referred to as a pump jack or pumping unit. This is the ubiquitous symbol of the oilfield that everyone sees when driving through the oilfields of Oklahoma or east Texas; it works like a teeter-totter. Pumping units come in a large range of sizes and styles and operate at a variety of speeds, depending on the depth of the well and the potential production rate of the well.

Bottomhole pumps are relatively common artificial lift methods as well, with an advantage of handling much larger production volumes than sucker rod pumps. Rather than lifting fluid periodically like bailing water from a boat, these pumps constantly push fluids up the hole. Like downhole motors for drilling, the pumps can be electrically or hydraulically driven. Gas lift is another effective artificial lift method to use when substantial amounts of gas are being produced in the field. In this case, gas is pumped down the annulus between the tubing and casing, usually in a very thin pipe, and injected into the tubing at various depths. The gas entering the tubing provides additional energy (from the pressure under which it was pumped in) and lightens the oil column, permitting the well to flow.

Secondary and Enhanced Oil Recovery • Regardless of the efforts to lift more oil out of the wellbore, every barrel of oil produced depressurizes the reservoir slightly, making it harder for the next barrel of oil to produce. Injecting some fluid into the reservoir to replace the produced oil is an obvious step, called secondary recovery. In the past, secondary recovery normally followed primary recovery, but petroleum engineers have seen that it is often advantageous to begin pressure maintenance very early in the life of a reservoir. Under the pressure of the reservoir, a certain amount of gas is generally dissolved in the oil, much like carbon dioxide is dissolved in soda pop. Like the carbonation in the soda, the gas dissolved under pressure tends to make the fluid flow from its container as the pressure is released. As the reservoir pressure declines below the bubblepoint (the pressure at which the first bubble of gas forms, as the boiling point is the temperature at which the first bubble of gas forms), increasing amounts of gas come out of solution in the reservoir. This causes two problems. One is that once enough gas has evolved to form a continuous phase in the reservoir, its lower viscosity permits it to flow much more readily than oil. The expansion of the gas in the reservoir is a large part of the mechanism that drives oil towards the low pressure point of the wellbore. As gas leaves, it deprives the reservoir of an important drive mechanism. Second, the gas flowing to the wellbore as a free phase occupies pore space that oil cannot use to flow, reducing the effective permeability to oil. Thus, it is desirable to maintain reservoir pressure above the bubblepoint, if possible.

Maintaining reservoir pressure requires injecting some fluid into the rock to replace the fluids withdrawn. The first and most obvious choice is water. Water floods involve converting producing wells to injection or drilling injection wells deliberately. The oldest and simplest (but generally least effective) form of water flood is to augment natural water influx. Since oil has filled a high point in a porous rock bed, water is often found in the same rock, below the oil. Although water is commonly referred to as *incompressible,* it is slightly compressed under the natural formation pressures. As the reservoir pressure declines, the water, often having a much larger volume than the oil, expands into the low pressure reservoir, pushing oil ahead of it. This process is called water influx or a natural water drive. The wells which penetrate the lowest portion of the oil reservoir are nearest the water-oil contact, and thus *water out* as the water moves in. Often, watered-out production wells are simply converted to injection at this point.

Studies have shown that an efficient water flood calls for deliberately planned well placement or flood patterns. Water moving through a reservoir always tends to move faster through more permeable streaks in the rock. This is called *fingering,* and when the water in a finger reaches a production well, it tends to dominate the production, since it is flowing through the most permeable intervals. Soon after the initial *water breakthrough,* a well is likely to produce so much water that it becomes too costly to handle for the amount of oil being produced. At this point the producing well has watered out and is removed from production, leaving a large amount of oil in the less permeable strata of the reservoir. The problem is less severe when the injection wells are closer to the producers; therefore, new injection wells are often drilled for a water flood.

No matter how carefully the well spacing and flood patterns are selected, waterflooding still leaves most of the oil behind (about 60% on the average). Water is less viscous than oil and tends to finger inevitably ahead of the oil. The means to address this problem is to make the water more viscous by the addition of polymers. However, even polymer-augmented waterfloods leave a large amount of oil behind.

The next level of enhanced recovery sophistication is miscible flooding. Here, a surfactant (surface active agent, like soap) is added to the injection water to take the oil up in an emulsion with the water, in which it will be carried to the production well. Miscible surfactant floods can achieve very high recoveries, approaching 95% of the oil in place, but are very expensive. Not only is the surfactant costly, but breaking the oil-water emulsion at the surface is costly too.

Another form of miscible flooding is miscible gas injection. If a large supply of natural gas is available, this can be a very efficient technique. Gas pumped into the reservoir at a pressure greater than the bubblepoint pressure can go into solution with the oil. As more gas is dissolved, the oil becomes less viscous and more compressible, thus it flows more easily to the production well. Of course, as gas goes into solution it also changes the composition of the oil, and hence the bubblepoint of the system. Furthermore, the system cannot reach a new equilibrium instantly as more gas is constantly injected; so the oil near the injector is very gassy, while that which is farthest away may not have experienced any change from its initial composition. In spite of these problems, miscible gas flooding can work very well. The author has worked on a field in Wyoming which was an extraordinarily successful example. The abundant gas was provided by the local utility company which needed a long-term gas storage project where they could keep a large supply that could be readily produced in the event of a supply shortage. They were happy to provide the oil company with gas to inject, free of charge, since they were not being charged for the use of the storage field. Everyone won—until the federal government saw what an eminently logical thing was being done and acted quickly to correct it. The government informed the utility company that since the gas was being used for a practical purpose, this could not be called a storage project, and the utility company would have to pay tax on the value of the gas stored, as if they had sold it. The oil company, having to buy the gas to inject, could not make money on the project and had to discontinue it, leaving behind a few million barrels of oil that would have been recovered had the project continued.

Gas can also be reinjected below the bubblepoint in an immiscible recovery process. The gas in such a case would normally be injected into the free gas cap. (When a reservoir is initially below the bubblepoint, it means that free bubbles of gas are present, which being lighter than the oil, migrate to the highest point in the reservoir, forming a free gas cap.) However, gas cap injection has an even more severe viscous fingering problem than water floods, because the gas is so mobile. Commonly, gas cap injection is not truly enhanced recovery but a more practical means of dealing with unnecessary gas production than flaring.

The viscosity of the oil is sometimes the biggest obstacle to recovery. Some heavy oils are quite literally tar and won't flow at all under prevailing reservoir temperatures. Raising the temperature thins the oil and lets it flow. A modification to water flooding is to heat the water before injecting it so that it heats the oil it contacts. Ideally, to carry the most heat, it is injected as steam, but it is very hard to mitigate heat losses in the wellbore sufficiently to get *live* steam to the reservoir. Hence, most efforts in this regard turn out to be hot water floods. They can be technically and economically successful.

Of course, if the point is to get a lot of heat into a reservoir, why not burn some of the oil in place? This is called *in situ combustion* or *fire flooding* and requires the injection of oxygen—and then lighting the oil. The oil that burns does greatly de-viscosify the oil near it and even evaporates some of it. Several firefloods were attempted during the oil price booms of the 1970s, but were generally difficult to control and economic failures.

Perhaps the most important fact to take from consideration of enhanced recovery is how much oil normally remains in the ground at the end of successful commercial field production. There is much more oil in the ground in known fields than the total cumulative world production. This probably represents an opportunity for as long as there is value to be gained from tapping petroleum resources.

Onshore and Offshore • Naturally, the first drilling for oil and gas occurred on dry land, where a rig could be easily erected. Indeed, even in the 1890s many rigs consisted of three cut trees lashed together at the top. The crown blocks were hung from the apex of the tripod thus formed. Of course, many more early oil wells were drilled from standard derricks, assembled on the spot, with the traditional image of criss-crossing beams. The structure could be either wood or steel—the wood derricks typically being left to stand sentinel over the well while the steel derricks would more often be disassembled, piece-by-piece, and reassembled over the next well. Some people are surprised, though, to learn that the oil industry moved offshore within its first half-century. By the turn of the century, the first drilling in the ocean was done from piers at Santa Barbara, California. In 1947, the first well distant from shore was drilled about 10 miles off the coast of Louisiana. Even as drilling operations moved farther from land, the early platforms were pier-like structures resting on the shallow bottom of the bayous. On these platforms, derricks were erected, and drilling commenced that continues in some of the same geologic *plays* into the end of the century.

Offshore drilling technology developed dramatically and

dominates much of modern-day exploration. As it was seen that petroleum reservoirs could be found in sediments under deeper ocean waters, it became impractical to build timber platforms from the sea floor to the surface. Drilling barges entered the scene. The derrick sat on a barge designed with an opening in its middle. (Think of it as being somewhat like a rectangular arrangement of pontoons.) The drilling rig straddles the opening in the middle of the barge. It was found that if the barge took on water to sink part way, it provided a more stable platform. This is the principal of *semi-submersible rigs,* which remain in common use at the writing of this book.

The popular image of modern offshore drilling and production is that of a platform of Brobdingnagian proportions. A large variety of platform types exists, with special designs for many conditions and innovations, and not all of them are large. However, a large platform can be the largest single investment in an offshore field; its design merits all of the attention it gets.

Platform installation is very expensive, so wildcatting is done from one of the following mobile marine rigs: drillship, barge, or jackup. If a discovery well and several confirmation wells prove that a reservoir contains enough oil (and/or gas) to pay off the huge investment of an offshore platform, then a platform is ordered and set. This fact helps explain why initiating production from an offshore exploratory play is likely to take 10 years or even more. The confirmation wells are only commenced once a successful wildcat has been drilled, and construction of the platform follows the successful confirmation well program.

Though there is a diversity of platform designs, only the general features of a platform will be considered. Commonly, a platform is designed for both drilling and production. One or two derricks may be mounted on the platform on tracks, which allow the drilling rigs to be skidded along a template (perhaps a slot in the platform floor) from one well position to the next. The combination platform also holds production facilities. Those facilities are used to separate oil, gas, and water; to monitor the production rates; dispose of unwanted fluids; and send the desired production either to a pipeline or storage to be accumulated for a tanker. Since the idea is to maximize the amount of the field which can be drilled from a single platform, most of the wells are drilled directionally, angling away from the platform to reach out into the field.

After drilling to a predetermined depth below the ocean floor, the well is *kicked off*—drilling at an angle from vertical. Until the 1970s, wells were often kicked off with *whipstocks*, which are massive, bent steel channels, which forced the bit in the direction the whipstock was bent. Angle can also be built by increasing the weight on the bit—the more weight, the more the bit tends to divert from vertical. Decreasing weight causes the bit to continue in a relatively straight line in the given direction. Decreasing weight even further causes the bit to tend to return to vertical. (Weight on the bit is controlled by the portion of the drillstring's weight the derrick holds. However, if additional weight is needed, heavy-walled pipes, called drill collars, are often used just above the bit.) A popular modern directional drilling technique is to use a bent sub and a downhole motor. The bent sub is simply a short length of pipe that is formed bent. The downhole motor turns the bit instead of the surface rotary table; it can be electrically powered or driven hydraulically by the mud being pumped through it.

When an appropriate number of wells is drilled to produce the field efficiently, the derrick may yet remain to service wells that develop leaks in the tubing or require stimulation, etc. The platform begins its production functions as soon as the first well is completed and placed on production.

OIL SHALE OR KEROGEN

Exploration • Huge resources of kerogen have already been identified by surface geologic observations in many regions of mature geologic investigation. These areas include Scotland, Canada, the western United States, Australia, Spain, and South Africa. Since kerogen can be thought of as oil and gas source rock, it is found in similar sedimentary basins, especially in paleo-deltaic and continental margin regions. (The paleogeography can be much different from current geography. For instance, the vast deposits in the Rocky Mountains are more than 1,000 miles from any existing oceans. However, they were deposited in ancient oceans.)

As with coal, rank and assay are important in defining the resource base. Samples of the rock in question are subjected to crushing and retorting, with the resulting shale oil yield measured in gallons per ton of original rock mass. The higher the yield, the better the resource. Of course, like coal, the shale's extent and depth of overburden must also be mapped. The level of North American demand during the 1970s led to the development of oil shale resources in western Colorado. The Colorado oil shales are vast, but some oil shale resources have even higher assays than the rich shales of the American Rockies.

Extraction • At the writing of this book, there is no active, commercial kerogen recovery process. There have been times when some notable exploitation of the resource has occurred. In the mid-19th century, in the United States, some oil shale development began but on a scale that was easily overwhelmed by the first oil production. The fact that Drake's well sputtered in at a mere 10 BOPD (compared to the 10,000 BOPD not uncommon in international oil exploration) gives an idea of how small the early oil shale industry really was. During the tenure of President Carter, the U.S. government launched the Synthetic Fuels Corporation and signed contracts with oil companies guaranteeing to buy a minimum amount of synthetic crude at a minimum price. Since these efforts started at the time the country faced its second oil shortage within a decade, and prices rising with no end in sight, the effort seemed reasonable, and so did minimum contractual prices of over $40/barrel. Four major oil company projects targeted the vast Green River Oil Shales of the western Rocky Mountains in southern Wyoming, Colorado, and Utah. Royal Dutch Shell and an Amoco-Oxy consortium folded their efforts in the face of technical difficulties even while prices were still climbing. Exxon abandoned their efforts in the early 1980s, cutting losses at several hundred million dollars invested in a pilot project that never produced. The Synthetic Fuels Corporation disbanded. Union Oil Company was the sole recalcitrant that pursued the effort to the point of bringing a pilot plant on line to produce synthetic crude. By the time that Union's (by then renamed UNOCAL) production commenced, free market oil prices had fallen well below half of the government-guaranteed $42.50 price. This much higher price for synthetic crude than natural crude helped to keep UNOCAL trying. The plant never sustained capacity production for more than a few days at a time and was abandoned within five years.[14] Therefore, oil shale production technologies will have to change markedly if the industry is ever revived.

Oil shales, when crushed and retorted, yield a small amount of liquid hydrocarbons which are comparable to crude oil and are hence referred to as synthetic crude. In addition to the energy losses in retorting, two significant problems challenge the technology. First, the spent shale residue occupies more volume, even after removing liquids and cooling, than the original rock. Since a ton of oil shale produces about 10 to 13 gallons of oil, a great deal of spent shale must be disposed of in a commercial-scale operation. Second, the crushing and retorting process mobilizes vast amounts of ultra-fine particles. Like coal particles, the dust can cause respiratory problems.

An environmentally intriguing concept was to attempt in situ retorting. In this method, wells would be drilled, a large cavern would be created, and shale fractured and rubblized explosively. Controlled amounts of air would be pumped in to sustain enough of a fire to retort the shale in place. Ideally, then, the liquid product would be produced through the wellbore just like crude oil. Unfortunately, the technology has never been able to fully address the problem of the swollen residue. Once the process began, swelling would quickly fill the cavern and plug off production and air to sustain the retort. This process never got beyond experimental and theoretical stages.

The alternative is to mine the material like coal but to add a retort step before transporting it. At Parachute Creek, Colorado, UNOCAL built a commercial-scale pilot plant and attempted underground mining. The retort consisted of a huge vat or tank into which crushed shale was introduced. The retort was rather like an enormous apple butter churn: a fire burned at the bottom of the vat, and paddles stirred the retort slurry. The process worked well on a bench scale and even on a pilot scale at their research facility in California. Unfortunately, the demonstration plant scale-up, begun in 1980, failed. The retort mixing was inefficient, paddles broke, the slurry thickened. Problem after problem beset the commercial plant, which produced a cumulative 4.5 million barrels of synthetic crude at a total construction investment of $654 million. This amounts to $145/barrel produced, well above even the highest price for which crude sold during the embargo years. The project continued to post annual losses through 1990 and was shut down in June of 1991.[15]

Even without the retort problems, the disposal problem would continue to hamper commercial-scale kerogen production. After spent shale is returned to abandoned portions of the mine, new places must be found for the additional residue. While the surplus awaits ultimate disposal, vast amounts of water must be sprayed on it to keep the mineral dust from blowing.

BIOMASS

Exploration and Fuelwood Deficits • Certainly, finding biomass requires little exploration. Direct observation works well here; one simply looks around, picks out some dead brush, a tree, or cow dropping, and goes after it. Yet many questions regarding the resource base merit a detailed and sophisticated examination. A substantial controversy continues to revolve around the question of how much biomass is produced that can be consumed for energy versus how much is actually consumed. No adequate data defines the actual consumption rates and growth rates in forests supplying firewood to Third World residents. More will be said of this in the chapter discussing the impacts of energy use.

Most biomass harvesting has been and continues to be done in the informal sector. This includes people collecting and cutting their own firewood as a *free good* and black marketeers gathering firewood for sale. The formal market, commercial production of firewood and charcoal, is very small. Since more than 1 billion people are believed to live in *firewood deficit* regions in 1980, with increasing deforestation, acquisition is a serious question.[16] (In spite of the controversy regarding the extent of the problem, the author has directly observed the lengthy firewood-collecting sorties of rural people in sub-Saharan Africa, and the exorbitant prices paid for the commodity by urban dwellers—attesting clearly to the fact that the deficit does exist.) To a North American, getting firewood may mean driving a pickup truck to the forest, firing up the chainsaw to cut a tree, and load the wood into the truck and bring it home. In some more environmentally benign cases, it means merely collecting prunings from your orchard or gathering other debris.

In lower income countries, the search for fuel is likely to mean a woman walking as far as 20 kilometers from her village to gather brush and woody material to carry home. Or it may mean picking up cow droppings, pressing them against a wall in the sun to dry out, and collecting them later to cook the evening's meal. One U.S. government official has suggested that he has observed people in India happily cooking with dung; however, he did not indicate that he had ever asked the privilege of joining in this rich custom.

People in urban areas, whether in industrialized or non-industrialized nations, cannot gather biomass as a free good. Rather they buy it from firewood or charcoal dealers. Since deforestation in lower income countries is a problem that has prompted laws against cutting any live growth, it is not difficult to imagine why much of the industry operates as a black market there. Often, the governments realize that people's need for energy to cook their meals—to meet their survival needs—supersedes any legal considerations and tolerate the black market. A government official in a southern African country drove with the author to visit illegal charcoal markets and discuss prices. Even though the very presence of a car gave away the government official, black marketeers chatted matter-of-factly with him about their prices.

Commonly, the biomass seller owns an old truck which he drives to the forests. Since wood contains a significant amount of moisture, which adds to the weight and bulk, the merchant often finds it advantageous to convert the product to charcoal. This is usually done by cutting down a tree, setting it on fire, and then covering it with dirt to smolder. Just as lighter organic compounds are driven off with the water in the coalification process, so are they in the charcoalification process. The process described is extraordinarily inefficient and probably encourages cutting live growth rather than searching for dead material to use.

More field research is needed to resolve such details of traditional and neotraditional biomass use. Kilns are more efficient than traditional charcoal production but more trouble, and portable ones would not be likely to handle entire trees at once. Alcohol or methane produced from biomass burns more cleanly and is more versatile. Production problems can be limiting factors for liquid and gaseous fuels derived from biomass.

In effect, the exploration phase for biomass energy production is not one of searching for where the stuff is to be found but rather searching for a viable means to use it. The notion that biomass can be utilized as a renewable resource, and some nostalgic ties to its use as *natural*, inspires the quest in parts of the industrialized world. Some people speak of firewood plantations as a means of providing biomass in a sustainable fashion. In this scenario, trees are planted to grow rapidly and produce good firewood. This has yet to be seen to work well, although the Department of Energy (DOE), created in the United States in response to the oil supply shortages of the 1970s, has supported research and pilot projects in wood plantations. Early estimates suggest that the risk of accidental injury may be 3 to 10 times greater in firewood plantations than underground coal mines—quite a disturbing comparison.[17]

The University of Rochester in upstate New York recently commissioned student intern studies of the potential for changing over the university's power plant from coal to biomass fuel. The idea was that wood fuel might be cleaner and cheaper in the long run, while its use would support a tree plantation industry in the region. The studies raised fundamental questions about the cleanliness issue, which will be discussed in more detail in the environmental chapter. A spring 1993 news release that the University had decided to switch to natural gas may give a pretty good idea of what the cost calculations revealed. It also stands to corroborate the notion put forth by Vaclav Smil in *Energy, Food, Environment*, that direct burning of biomass is an ill-advised technological step backward.

Biomass Agriculture • In the University of Rochester's preliminary studies, Short Rotation Intensive Culture (SRIC) was selected as the more cost-effective wood acquisition approach, as opposed to waste-wood gathering or traditional (long rotation) forestry. Waste-wood utilization was deemed ineffective because of potential supply shortages; the supplier provides woody material as a secondary (or tertiary) enterprise, without primary contractual obligations. Long rotation forestry was deemed ineffective because it is less productive on a mass per acre per year basis.

Intensive is a good word for SRIC: as many as 15,000 trees are planted per hectare (1 hectare = 2.47 acres). Fast growing poplar and willow are favored species for the plantations. A traditional eastern United States hardwood forest might have up to 3,500 trees per hectare. (Six-foot spacing would make a fairly dense forest.) Indeed, new technologies seek even greater intensiveness. The ultrashort rotation system, also called the wood grass system, plants 16,000 to 180,000 trees/hectare on a 1.5 to .5 foot spacing—as much as a tree every 6 inches. Certainly, these trees can never mature; they don't have room. Harvesting is conducted every 3-10 years in the conventional SRIC plantation and every 1-2 years in the wood grass variation. Some of the technological problems with firewood plantations involve ecological problems. Typically, a single or small range of fast-growing carbon-rich trees is selected to plant.[18] Even as Rachel Carson noted in her seminal *Silent Spring,* in 1962, considerable risk is associated with establishing a lack of genetic diversity. She observed that the popularity of elm trees for shade caused many suburban streets to be lined almost exclusively with elm, which provided beauty, uniformity, and fertile ground for the spread of Dutch Elm Disease that succeeded in virtually eradicating that grand species in North America. Some agricultural scientists are now writing about similar risks for farmers. Will it be any less so for fuel wood plantations? Perhaps it is impractical to try to tie the industrialized world's lust for energy to reafforestation schemes. Natural forests have a wide range of species that provide robustness to the ecosystem and nuisances to agricultural efforts.

Agro-forestry methods might have more appeal. In these schemes, tree planting is intermixed with agricultural crops, sometimes as tree rows between cultivated plots and sometimes as trees planted in the midst of the crops to provide shade as well as wood growth. This at least provides some diversity and can help with erosion and provide wind breaks. Of course, this method probably will not work in the large, commercial farms of Kansas. The plan lends itself most readily to small subsistence-type farms, where the plots are small, heavy machinery is not utilized, and the energy demand is largely domestic. Some authors refer to the potential advantages of using fruit or nut trees in the planting, which would provide food, perhaps even a cash crop, and prunings for firewood. Studies in Africa suggest that agro-forestry could be cost-effective in terms of increased agricultural productivity and mitigated erosional effects.[19] Of course, trees take time to grow, and it is difficult for people living on a subsistence level to allocate time and energy to such long-term investments. Nonetheless, agro-forestry is designed to have its best applications in the small farm setting typical of subsistence farms and increasingly atypical of commercial North American farming.

Large-scale commercial farm production of biomass feedstocks to make fluid fuels is another approach. An ethanol fuel industry can be established simply by distilling more grain to make more alcohol in a manner very similar to liquor production (but without the FDA constraints).This strategy is appealing at face value because of the massive production rates of modern agriculture. It is true that industrialized nations have established remarkable agricultural productivity per worker hour or per acre. With less people employed on the farms and less acreage in production, the United States produces more food crops than ever before. Some argue that this agricultural production provides the renewable energy resource base of the future. But American agriculture has flourished with massive energy subsidies.The subsidies refer to the input of inexpensive fossil fuels into farming. The fossil fuel input is not only seen in the form of energy to run tractors but in the conversion of fossil fuels to ammonia for fertilizer. These inputs have been instrumental in the dramatic increases in agricultural productivity. It can be persuasively argued that the *so called* subsidies do not make modern agriculture inefficient in producing high quality food. The subsidies do seem important in assessing the potential for agriculture to produce alternative sources efficiently.[20]

These subsidies are not direct government gifts; rather, they are subsidies of energy purchased by the agricultural businesses to till more land with less labor, to add more fertilizer, and irrigate more. The subsidy reflects the fact that land and wages have become economically dominant over other cost inputs, and the fact that market forces (some would argue market distortions) have greatly favored very large farms, where the large scale application of fertilizers, pesticides, herbicides, and irrigation can be conducted. But how sustainable is an energy production system that is more than 50% energy intensive (i.e., burns one Btu for every two it produces)? Trends would suggest that agricultural energy intensiveness is likely to grow as more and more is produced from the same, and possibly dwindling, lands. It is true that the United States currently has surplus corn production. Is it possible to use that surplus to produce ethanol to fuel vehicles? It probably is. But factoring in the alleged renewability at the same time as taking advantage of a pre-existing surplus is a fool's paradise. The ability to draw on an existing stockpile is a depletion process, not a process of making use of resource renewal rates. The long-term benefits of agricultural ethanol production must stand on their own merits, which look questionable at the writing of this book.

Some humanists would indeed argue that it is unethical to refer even now to a food surplus at a time when a quarter of the world's people suffer from malnutrition. In this argument, if we gain adequate global perspective, we should see the question as whether to burn grain alcohol to drive to the mall or use the grain to feed starving children. This is an interesting ethical question beyond the scope of this book.

Biogas • Another means of using biomass is in the creation of methane through anaerobic digestion of organic material. This process emulates the early stages of the evolution of petroleum and coal. The organic matter most often utilized in biogas production is fecal material. Two great advantages that biogas generation has are 1) the solid residue is as useful for fertilizer purposes as the original manure, and 2) the product is clean-burning methane. Approximately one pound of organic matter can yield eight cubic feet of methane gas.[21]

Wherever a concentration of living creatures is producing fecal material that can be readily accumulated, it can be placed into a digestor to create a *free* supply of methane. The digestor is a tank, much like a septic tank, but more complexly sealed. In the digestor it must be possible to introduce the waste without introducing significant amounts of oxygen or permitting significant amounts of methane to escape, if the digestor is to be in continuous use. A pipe coming off the top of the digestor provides methane to appliances.

This author would suggest that biogas generation has real potential wherever large amounts of organic waste are present in a concentrated locale. For instance, more energy would probably be consumed gathering dung on a free-range ranch or farm than the dung would contain. However, a feed-lot or chicken coop has the material in a concentrated location. Human excrement works too. Indeed, for many years, sewage treatment plants have collected methane to provide on-site energy that was spontaneously generated from the waste containment facilities. Generally, biogas generation can be viewed as a win-win situation. In its most efficient form, it utilizes concentrated waste products that must be handled anyway and produces clean energy and a residue that is highly valuable as fertilizer. This author has not yet found data on the additional resource base of biogas, but would speculate that it is most practical in rural and developing country applications. (Many urban areas in lower income countries lack sewage gathering lines and treatment facilities. Could the provision of energy through intentionally designed digestors at treatment plants offer some economic incentive for installing the much-needed facilities?)

Energy from Waste • Some projects to derive commercial energy from municipal waste already exist in a number of forms. The questions pertinent to this chapter relate to gathering the material and preparing it for combustion. The most obvious approach is direct burn of collected garbage. In this, a cursory sorting removes metals and large noncombustibles from the refuse stream, and the remainder enters a combustion chamber. In some forms of the technology, the product is more thoroughly segregated by a series of grates, and then a large hammer flails and shreds the combustible material. Then the resultant can be pressed into pellets. The advantage to the more involved process is the production of more uniform fuels with which furnaces can operate more efficiently. (The combustion process will be discussed in detail in the conversion and end-use chapter.) In the United States, 14% of municipal waste was burned for heat in 1990 while Switzerland burned 74%.[22]

One form of solid waste that is routinely utilized even in the United States is sawdust and woodchips. Most sawmills burn their by-product for process heat. Some compress the sawdust into pellets to be sold for specially designed, high output home stoves. This is one of the few utilizations of waste burn that does not meet with opposition based on environmental grounds. The preferred method of disposal in industrialized countries is burial in landfills. So, is landfill disposal environmentally superior to burning wastes for useful heat? This debate will be explored in Chapter 5.

Incineration of solid waste can be considered as separate from other forms of waste burning, in that it implies that additional (commercial) fuels are used to ensure a controlled combustion at a specified temperature. This may be an approach to address environmental concerns, although many people oppose waste incineration, still under the environmental banner. Industrialized countries could certainly find an energy source close at hand through controlled incineration of wastes, including combustible wastes that are currently dubbed *hazardous.*

NONCOMBUSTION SOURCES

SOLAR POWER

Exploration • As with biomass, solar energy exploration is a moot issue. Exploration consists of compiling readily observed data on insolation. The data should include the large and small insolation cycles and some standard deviation data. A view of the amount of insolation available throughout the world is shown on Figure 2.5. It is not only important how much annual insolation a place receives, but how the rate of insolation compares to peak demand cycles. For example, while southern California receives high annual insolation, winter is the rainy season, with less insolation than summer; this impairs (though it certainly does not eradicate) the solar heating potential in California. Solar heating generally requires some back-up system in this case, as will be discussed in the conversion and end-use chapter. In addition to seasonal cycles, since day and night cycles affect the availability of solar energy, storage design is also necessary.

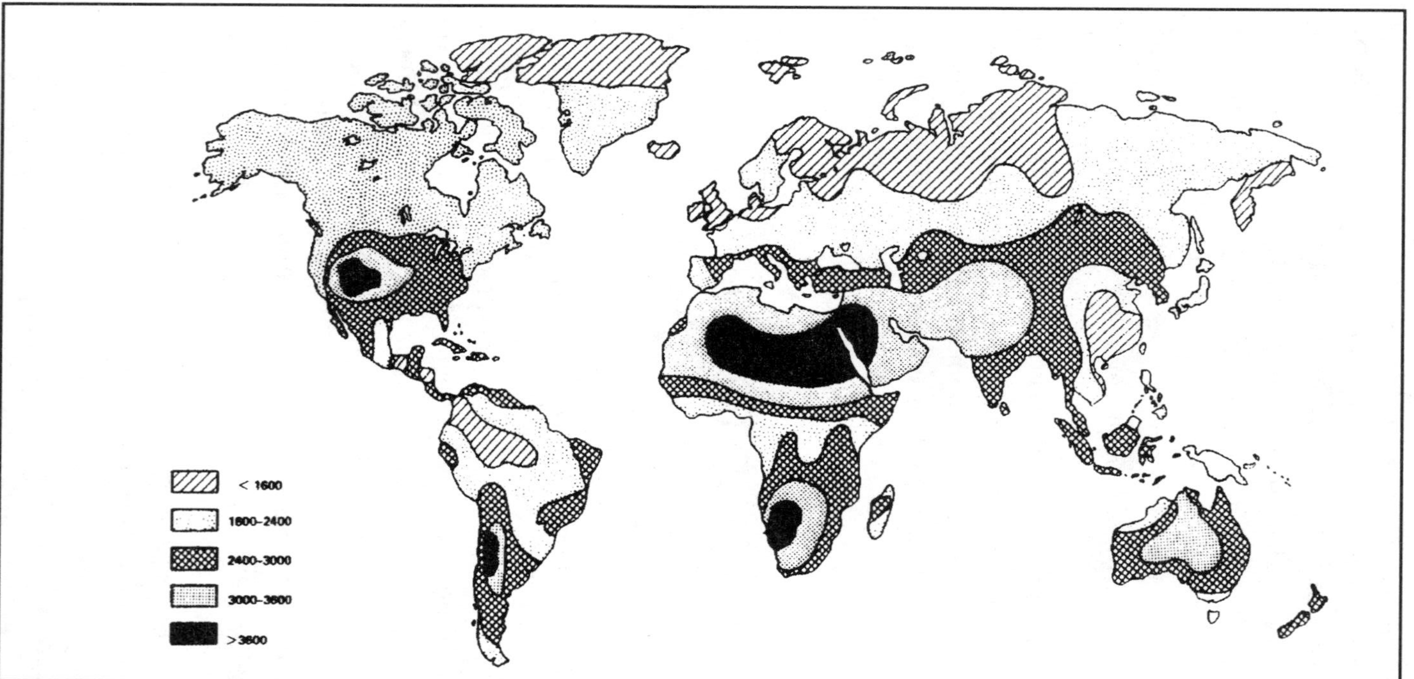

Figure 2–5 Global insolation.
The shaded areas on the map indicate regions with high annual sunlight.
Courtesy Mitre Corporation, from *Global Energy: Assessing the Future.*

Production technologies are the issues to be considered in this chapter. First, the distinction between active and passive systems is important. Active systems involve some designed human intervention that may use machines to transfer energy from the collection point to some storage or utilization point or some technology to convert the light energy into some other form (generally electrical energy). Passive systems are simply designed as collectors to take advantage of the light/heat of incident solar energy.

Active Solar • Electricity is doubtless the most convenient and versatile form of energy for human use. Consequently, conversion of solar energy into electricity is a much desired process. Photovoltaic and solar thermal electric conversion are two dominant means of accomplishing this.

Some technologies (particularly photovoltaic) can make use of diffuse sunlight, while others require concentrated rays. Technologies requiring concentrated light are rendered impotent on cloudy or hazy days and in the early morning and late evenings, while those able to convert diffuse rays are not limited in this way. This makes insolation charts not easily applicable to both cases. Direct and indirect insolation must be differentiated.

In general, the closer to the equator, the more dependable insolation is from season to season. The drier the climate, the more insolation reaches the earth on a day-to-day basis. Efforts to acquire solar energy must seek areas which optimize insolation and demand. The demand profile must match the insolation profile adequately. The type of solar energy project must be designed with the regional characteristics of both insolation and demand in view. Most solar thermal electric conversion (STEC) projects would not be appropriate for humid tropical locales; in spite of sizable gross insolation, the sunlight is often diffused by clouds and haze.

Diffusion has inspired the search for more intense sources of solar radiation. Since the atmosphere is the primary source of diffusion, placing solar collectors above the atmosphere seems like a logical solution. One possibility is to place satellites designed with huge solar collectors in geostationary orbit (constantly over the same point on the earth). The satellite would transform the gathered light into microwave radiation and beam it to a receiving dish on the surface. While it is true that a satellite would be exposed to a great deal more solar radiation than any ground station, and that radiation in the microwave frequency travels relatively undiffused through the atmosphere, this acquisition scheme does not seem particularly hopeful for the near future (i.e., through the first half of the 21st century). The satellite itself is very expensive. In addition, most proposed descriptions suggest that the collector consist of a photovoltaic cell array, which is still constrained by the low efficiencies of photovoltaic conversion. The microwave receiving and electricity generating station would also be expensive. At least at face value, these satellites seem a bit risky as well. In the satellite collector arrangement, the satellite must do the tracking and aim its beam, rather than being the target of beams aimed from ground stations. If the satellite-aiming system suffered a failure, could intense radiation in the microwave band be beamed into a populated area? Could this present a risk of catastrophic failure, comparable to loss of containment in a nuclear plant?

Solar ponds offer a very inexpensive method of harvesting energy from the sun. Salinity gradients naturally occur in extremely salty waters because water with a higher salt content is denser. Therefore, the water at the bottom of a salt-saturated pond does not rise as it normally would through convection when it takes on heat from the bottom of the pond. The sun striking the bottom of a pond heats the bottom and the heat is transferred to the water on the bottom, which normally rises just as hot air rises in a room. Because of the salinity gradient, though, the water stays on the bottom of the pond, taking on more and more heat, and creating a large temperature gradient between the surface and bottom waters. This temperature difference can be exploited to evaporate a carrier fluid and drive a turbine. The efficiency is very low, but since it can be very cheap, it may have application in isolated areas. Even with only a 2% conversion efficiency, a pond 1,000 feet by 1,000 feet could generate a megawatt of power if the insolation reaches as high as 500 watts per square meter, as it does in sunny regions.[23]

Passive Solar • Technically, while humans experience heat and light as two different phenomena, visible light and heat are simply different parts of the electromagnetic radiation spectrum. (As this name suggests, then, so is electricity, but solar radiation reaches the earth primarily in the light and heat ranges of the spectrum.) Passive solar technologies then make direct use of heat and light. A popular form of demand in the industrialized world is space heating, and passive solar designs can play a significant role in this area, especially where there is considerable sunlight during peak heating demand times.

One of the oldest and most prevalent forms of solar passive design is in space heating and lighting. Fenestration, the placement of windows and doors in buildings, goes back to the earliest structures. While people desired protection from the elements, a means of admitting the light and warmth of the sun was comparably important. The Golden Age Greeks in many cities developed solar architecture; buildings were almost universally oriented to southern exposures. The houses took advantage of winter sun coming in at a low angle, while roof overhangs shielded the windows from the intense summer sun. Into the middle ages, fenestra were merely openings in many areas. Doors or shutters could close against the weather and intruders. Mica, a naturally planar, transparent mineral, probably provided the first solar gain windows, offering a barrier to air movement while still admitting solar radiation. Glass, originally discovered in the Middle East (perhaps in Syria in the 4th or 5th century B.C.), led to the introduction of glazed windows during the Imperial Roman period, and ultimately evolved into the strong, clear windows we know today.[24] Into the latter half of the 20th century, windows still permitted significant heat loss by conducting interior heat too efficiently to the exterior, where air movement could carry the heat away. Double panes of glass (double glazing) were introduced as a solution. The ongoing efforts to design windows that optimize solar gain while minimizing heat loss are central to passive solar developments. Some experimenters, such as Amory and Hunter Lovins at the Rocky Mountain Institute, through a combination of passive solar and conservation measures, claim to have realized greater than 90% solar heating. This is an area with substantial potential in most industrialized countries, where space heating is typically a major household energy demand item, and high quality building materials are readily available.

Closely related to space heating is passive solar water heating, in which water pipes are simply exposed to the sun to gain energy. If 10% of all American water heating were conducted by solar methods, at 1990 levels of demand, 1.5% of the total household energy demand could be met.[25] Passive solar heating design requirements on all new American construction could account for only a very modest faction of total energy demand, but a non-negligible quantity of fuel saved. Naturally, these designs also call for substantial insulation/conservation measures.

In most lower income countries, space heating is not a primary demand, and in those places where it is, the unavailability of efficient building materials jeopardizes the potential for passive solar heating. Cooking is the primary domestic energy demand in most lower income contexts. Several groups have tried to introduce solar cookers to meet this demand, with a noteworthy lack of success. In some cases, the technical design is faulty, or the human component is inadequately considered. The chapter on conversion and end-use will explore the importance of designing to meet needs in their human context and the fallacies of technical design that may be encountered, especially in the solar arena. For the scope of this chapter, suffice it say that the exploration for solar applications is constrained in the area of cooking by 1) the time regimes within which cooking must be accomplished, 2) the inelasticity of demand, 3) the immediacy of the task, and 4) the difficulty of supplementing cooking energy.

The solar acquisition strategy that truly requires no significantly new technological developments is passive heat gain in space and water heating.

WIND POWER

Low intensity, continuous-demand tasks such as the following have long utilized wind: milling grain, pumping water from wells to the surface or from point to point, and propelling marine craft. The quest to expand wind-energy acquisition has moved naturally to generating electricity. Indeed, the move toward electricity is natural, not only from the consumer's perspective of generating the most versatile and desired energy form but on technical grounds as well. The rotary movement of windmills is similar to that of electric generators. However, generator dynamos require high speeds of revolution which are not typical of traditional windmill designs. The efficiency of a wind electric turbine is a function of the height of the tower and of the cube of wind velocity. The velocity cubed term is the real problem; it means that doubling wind speed improves the electric output by a factor of 8, further indicating that very high average wind speeds are best for generating electricity.

Wind is a global phenomenon, but prevailing speeds are functions of topography and climate. The distribution of wind velocities around the world is illustrated on Figure 2.6. The dark areas, indicative of high wind intensity, are generally the best technical prospects for siting wind turbines. A few areas of high population density corresponding to high wind intensities have been highlighted as an indication of a first step in matching demand with resource availability. As with solar energy, another exploration step in utilizing wind involves comparing seasonal and daily wind patterns with demand patterns.

HYDROPOWER

Exploration • Conventional forms of hydropower that utilize the conversion of potential to kinetic energy in falling water have very straightforward exploration, which consists of simply addressing the question, "Where does a sufficient head and flow rate exist?" As

Figure 2–6 Global wind velocities.
The shaded regions have high average wind velocities, making them most desirable for siting wind power facilities.
Courtesy Mitre Corporation, from *Global Energy: Assessing the Future.*

environmental concerns dictate against the large-scale hydroelectric dam projects that prevailed in the middle of the 20th century, attention is turning to small-scale applications. In general, wherever a stream or river is moving quickly, the potential exists. In a sense, this opens the door to finding many new hydropower resources, but it also signifies a return to the scale of centuries ago. Perhaps in the industrialized world, some exploration efforts can look to the old watermills that have been abandoned in the times of cheap, abundant energy. Some international development agencies are seeking such small-scale sites in nonindustrialized nations. While the smaller applications do not require the dramatic heads and flow rates required to turn the massive turbines at acceptable rates, the power produced is still a function of these two factors. It is not likely that the world will see many new, large dam projects. Even if many small projects are implemented, the total provision of energy is likely to be modest.

A primary constraint on hydropower utilization has been the difficulty of transporting the energy large distances. The mechanical power produced by a water mill must be used on site. However, hydroelectricity is not only more versatile but also more mobile, though it still suffers significant efficiency losses with distance of transport. One solution to this problem is ultrahigh voltage systems. If superconductors develop dramatically, they could greatly minimize electric transmission losses and open hydro potential in much more remote areas. For instance, in North America, a very large proportion of the known, untapped potential for this resource is in Alaska and northern Canada, but transport distances are comparable to transoceanic transport. If very efficient transmission becomes possible, these resources could become reserves.

Acquisition • The kinetic energy of moving water must be captured in a mechanical energy that can be converted either into useful work or into an intermediate energy carrier, which is most commonly electricity. The earliest efforts to make use of the energy of flowing water are not substantially different in principle than modern hydroelectric technologies.

Somewhat more than 2,000 years ago, the first known water wheel, the *Noria,* was invented. It consisted of pails attached to the perimeter of a wheel mounted over a stream. As the bottom-most pail filled, the force of the moving water turned the wheel; bringing the next pail down and lifting the pails to the top to dump into a flume or channel to flow to the point of use. This invention was followed shortly by the first geared mill, the *Vitruvian.* The gears permitted changing the axis and speed of rotation. In fact, millers are reported to have resisted the introduction of water mills because they did not want to move their operations to streams and rivers. The effectiveness of the new energy source won out, though, and by the 6th century A.D., water mills appeared throughout what is now western Europe.[26]

Effectively, the evolution from water mill to hydroelectric power plant was straightforward. The same sort of wheel, with blades that are turned when struck by water, produces rotational movement. Whether that motion is used to turn a grist wheel or to spin a turbine is essentially the same process. Water must be directed through a channel (sluice) which holds the wheel or turbine to ensure that the water's force is delivered to the device, rather than diffused around it. Commonly, a dam is constructed to confine a portion of the moving water and provide a large head across the wheel or turbine. The water head (height of the water column) determines the pressure or force per unit area exerted by the water at the bottom of its fall across the device blades. Even in rapidly moving water, the head that can be converted to mechanical energy is only the height difference from immediately above the device to immediately below it. The dam concentrates the height change of a great length of the river at one point.

Wave and Tidal Energy • A great deal of energy flux can be seen in waves and tides. Exploration for the resource, again, simply requires direct observation. For both wave and tidal energy, the potential sites are necessarily coastal, and a number of prime sites are well known. Weather conditions and slope of the continental shelf and the coast determine average wave activity; however, there is no need to deduce potential from these factors when the actual resource can be observed directly. The question lies in the means of acquiring energy from these systems.

Tidal power can use the same technologies as conventional hydropower. In fact, the Dutch employed tidal energy centuries ago. Again, dams and sluices are used to collect and direct a portion of the water through a waterwheel or turbine. As tides rise, they fill a sluice (a channel in the tidal flat). The water is trapped in the sluice and forced to drain back to the low tide ocean level through a turbine. The dramatic Bay of Fundy in Nova Scotia, Canada, with its remarkable tidal fluxes, produces a very large head. The longest-running tidal hydroelectric plant has been the French Rance project.

Wave power cannot be readily captured behind a dam, so it requires the development of new technologies. Since electric generation is the desired carrier form of energy, efforts focus on means to convert the bobbing motion of waves to a rotational movement. One approach involves fluid chambers in a buoy. As waves lift and drop the buoy, fluid is moved from a higher to lower chamber across a small turbine.

GEOTHERMAL POWER

Exploration • Since geothermal energy exists in the ground everywhere, exploration aims at finding sources where abnormally high geothermal temperatures are near the surface. Direct observation led to the discovery of much of the geothermal energy already exploited. Hot springs, fumaroles, geysers, and volcanoes offer evidence of magmatic intrusions into the upper crust of the earth, carrying immense amounts of heat with them. Geothermal exploratory drilling seeks to find porous and permeable aquifers near enough to the subsurface heat source to have taken up a good deal of the heat, while being close enough to the surface to be reached and produced efficiently. The early stages of exploratory drilling often only penetrate a few hundred feet to establish the near-surface geothermal gradient. The drilling and completion technologies are borrowed from the oil field. However, some design modifications are necessary to account for the high temperatures encountered, often in excess of 400°F. Special *hostile environment* logging tools are required because these temperatures can melt the circuitry.

Production • Well completions require some special attention. The temperatures cause steel to expand; a 3,000 ft geothermal well, even though the percentage expansion is small, can push the top joint of pipe several feet out of the ground during production. (Pre-expanding the pipe is generally not advisable, because it can cool and contract during off-production cycles. It would be embarrassing to have to report that your wellhead was sucked into the ground.) Expansion loops and legs are the typical solution to this problem. These are placed not only at the well head but on the field production line as well. Roller coaster loops of pipe can be seen; these are able to expand and contract without pushing or pulling against fixed points.

The steam produced is carried to a turbine which generates electricity. Concern for heat loss is the major limiting factor in geothermal exploration and production. Inevitably, heat is lost through the walls of wellbore and surface tubular goods. Insulation helps but can't be perfect. Thus, the depth of the well is limited by the issue of heat loss, as well as the distance between new wells and a geothermal plant. Overcoming this obstacle requires the development of *superinsulators.*

Open and Closed Loop Production Systems • Since the produced steam is likely to contain contaminants, and since sustained production from a geothermal reservoir requires maintaining reservoir pressure, produced fluids are normally reinjected. A variation that greatly mitigates the potential for escape of dissolved contaminants in the geothermal brine is to create a heat exchanger by pumping a volatile fluid through the hot water. This can be done underground and can expand the potential of exploiting reservoirs in which the brine is not hot enough to be *high quality* steam.

Types of Geothermal Reservoirs • The other limiting factor in standard geothermal development is finding a porous and permeable aquifer (geothermal reservoir) in the vicinity of the subsurface heat source. Recent work focuses on a potential means to get around this obstacle by tapping a different kind of reservoir. This is called the Hot Dry Rock Geothermal Project. It involves fracturing a bed of hot rock and injecting water, or some other carrier fluid, and then producing the injected fluid. If it is successful, it would permit significant growth of the geothermal industry. It looks hopeful in that it builds on successful existing technologies. Fracturing technologies have seen considerable growth in the oil field. Certainly, injecting water into a porous and permeable formation is feasible, but injection into impermeable rock is tougher. Presently, concerns about adequate fracture connection and control are prevalent. It is essential that fractures be well interconnected so that permeability is created. An ideal technology would be to inject water into one well and produce it from another, emulating water flooding technologies of the oil field. This approach not only permits a continuous pressure gradient to be established and maintained, moving fluids in the desired direction, but brings the injected fluids into contact with a large portion of the hot rock reservoir, with significant residence time facilitating the heat transfer. In order for injection from one well to reach another, the permeable paths must be well connected. Furthermore, they must be contained so that the injected fluids are not lost in fissures, carried off to other formations or parts of the field.

These constraints seem addressable by improvements to existing technology. The Hot Dry Rock Project is able to investigate one form of stimulation that has been dismissed in oil and gas applications as hopelessly impractical: fracturing the formation with nuclear explosions. Project Plowshare is the geothermal variation on the ill-fated hydrocarbon venture Project Gas-Buggy. The Plowshare idea has an advantage over Gas Buggy in that the radioactivity can be contained in the subsurface to a much larger degree. The well can be produced with a heat exchanger fluid, keeping the radioactive brines enclosed. Even the exchange fluid should be kept in a closed system so that they do not escape to the atmosphere. The idea of nuclear fracturing is that the incredible forces actually liquefy and vaporize some rock, creating a cavity near the wellbore. Meanwhile, the shock wave spreads through the reservoir, fracturing it. The technology seems to have some promise, but the intense heat is capable of glazing rock at the cavity face, creating an impermeable barrier. Controlling the fracture in Project Plowshare is not as essential as in an oil or gas operation because Plowshare seeks to produce heat rather than a specific formation fluid.

Ocean Thermal Electric Conversion (OTEC) • OTEC utilizes the temperature differences between surface and subsurface similarly to geothermal projects. In the OTEC case, ambient temperatures at the surface vaporize a fluid that is then pumped down tubing extending into the ocean to be condensed by the cool subsea temperatures. The ocean is used as a heat exchanger. Charting currents is an important first step towards identifying regions in which the hydrothermal gradient is especially steep. Like geothermal projects, heat losses in the tubing are a major limiting factor. Thus, it is essential to find regions where cold subsurface waters are found at relatively shallow depths.

NUCLEAR POWER

Currently, all commercial (peaceful) nuclear power applications employ the energy released by the fission (splitting) of very heavy atoms which are generally extracted from uranium and/or thorium ores. The best known and first discovered uranium ore is pitchblende, which can have uranium dioxide concentrations of 55 to 75% and uranium trioxide concentrations as high as 30%, making a very rich ore. Generally, the geologic formations from which uranium is produced are sandstone deposits, quartz-pebble conglomerates, proterozoic-unconformity-related deposits, disseminated deposits, and veins—in descending order of estimated total resource contributions.

The sandstone deposits are commonly of fluvial (river) or near-coastal marine origin. These deposits can be associated with hydrocarbon resources, as they are normally found with biotic residues and require a nonoxidizing environment. Oxidized uranium compounds are mobile because they are water soluble, while reduced compounds are not and precipitate from solution. A reducing (acidic) environment is therefore ideal. Hydrogen sulfide, which may be produced from the anaerobic decomposition of organic material, can provide the reducing material.

The quartz pebble conglomerates appear to be extensive but typically contain low-grade ores of uraninite. They are commonly of lower Proterozoic age (about a billion years old), and the uranium-bearing compounds are found in the interstices of the rock grains.

The Proterozoic unconformity-related deposits are most often found in lower Proterozoic rocks of 1.5 to 1.8 billion years of age. They are found just a little below a broadly occurring unconformity of that age. An unconformity refers to a contact between rocks of significantly different age, most commonly due to erosion of intermediate age rocks. These tend to be the richest known deposits, with high grade ore and extensive accumulations. Exploration is conducted in Pre-Cambrian shield regions, where these old formations are likely to be found relatively near the surface.

The disseminated deposits are found in igneous and metamorphic rocks including granite and its metamorphic descendants, schist and occasionally gneiss. The uranium is believed to have been deposited by fluids moved along fractures and bedding planes. Erosion due to weathering is a probable important aspect of the deposit becoming rich enough to qualify as an ore.[27]

While other types of deposits are known, and some researchers have considered even the possibility of extracting trace quantities of uranium salts from the oceans, the preceding represent the dominant known resource base. If nuclear power grows in significance, breeder reactors or the advent of controlled fusion would most likely obviate extending the search for uranium beyond the traditional sources. Clearly, the formations in which uranium can be found are geologically diverse, including hydrocarbon related sedimentary rocks, igneous, and metamorphic—no rock type is excluded. However, the geologists often narrow the search to areas seen to have a majority of known commercial veins: sedimentary basins like those sought for oil and gas, as well as to areas of very old rocks at the surface, and to areas having granitic and post-granitic rocks.

The principal exploration tool for uranium has classically been the Geiger counter. Properly called a Geiger-Müller counter, it consists of a gas-filled tube with an anode and high voltage electrical potential. When a gamma ray strikes an atom of the gas, it imparts enough energy to an electron to free it from the orbital. In the presence of an electrical field, in the gas-filled chamber, the ionized gas particle is drawn to the anode, causing a surge in the electrical current, which is recorded. In the presence of uranium, the gamma rays emitted by the natural radioactive decay process will cause high readings on the Geiger counter.

The uranium deposit may be exploited by mining or in situ leaching. The mining is typically of the underground chamber type, similar to that described in the coal section of this chapter. Miners face similar dangers as coal miners, but the energy intensity of nuclear fuels is so high that far fewer workers are needed to extract far smaller amounts of ore than coal to produce the same amount of energy.

In the case of traditional, mechanical mining techniques, leaching is normally done on the surface after extraction. In-place leaching is the compromise process which offers an exception to this observation. In-place leaching involves stope-type mining. After caving the material in, the leaching agent is pumped through it and the uranium salts extracted.

The in situ leaching process leaches the ore in the same process as extracting it. The process is very similar to the flooding processes of enhanced oil recovery. Fluids are pumped into injection wells and withdrawn from production wells, using similar spacing patterns as those developed in the oil patch. The fluid injected contains leaching agents. An oxidizing agent (e.g., ammonia, sodium bicarbonate, or hydrogen peroxide) is applied to reduce the uranium ions from a +2 to a +4 valence state, thereby making the uranium more soluble and hence mobile. In some cases, the oxidizing agent itself dissolves the uranium; for example, sodium bicarbonate mobilizes urano-carbonate salts. In other cases, sulfuric acid chases the oxidizing agent, carrying urano-sulfate salts to the production well. The uranium-bearing *pregnant* fluids are passed through ion exchangers, which remove the uranium and pass on a *barren* effluent, which has more leaching agents added for reinjection. The process of in situ leaching has found its use exclusively in sandstone deposits. These deposits have some natural, continuous permeability which allows fluids to flow from injector to producer while contacting a large portion of the deposit.[28]

It would seem logical that the igneous and metamorphic formations would be less susceptible to efficient flooding processes, because these rock types do not have pore spaces that create naturally permeable channels as do the granular sedimentary rocks. While fracturing has been considered to extend the application of in situ leaching, the nature of artificially induced fracturing is such that it would be difficult to ensure that the fractures contact a very large portion of the ore body.

The chapter which discusses the impact of energy use will consider health effects of miners to some extent; however, it is not a concern for the miners that motivates most of the antagonism towards nuclear power.

CONSERVATION

EXPLORATION

Once again, exploration may seem to be a trivial category for the energy resource of conservation, yet the subjectivity of conservation potential lends this category some significance. Two criteria should be applied when assessing conservation potential. First, what is the cost of not using the energy to be conserved? Second, how significant is a specific form of conservation? Not using energy, by foregoing the activity fueled, is the simplest form of conservation. The cost of not using energy is equivalent to the loss of benefit derived from consuming it. When energy is used to meet fundamental needs—cooking food, providing survival heat to a personal shelter—the cost is clear and overwhelming. Thus, only discretionary energy use can be addressed in the abstinence form of conservation. The problem that arises when speaking of conservation in the context of the nonindustrialized world is that even nondiscretionary needs are not being adequately met in many regions. Thus, many would argue that total energy consumption must increase in these regions to provide a reasonable quality of life for the region's inhabitants.

The second criterion for conservation efforts to have optimal yields indicates that they should focus on relatively high demand areas, where the net savings potential is high. For example, while one could argue that the energy consumed in road races or monster truck exhibitions is purely discretionary, the potential for savings is minimal and perhaps not worth a political campaign to ban such events, which obviously entertain a significant number of people.

Another form of conservation is improved efficiency. It is more broadly applicable because it does not necessitate any sacrifice in lifestyle. This requires technologic innovation in some cases or capital allocation in many other cases. Money can be (and is) assigned to research into improved materials and conversion processes to maximize efficiency. Financial resources can also be applied to employing more efficient options. Improved insulation in homes and heat recovery processes in industry are two examples of this form of conservation.

Improved efficiency is a potentially viable form of conservation for Third World applications. A modern, wood-burning stove can easily be four times more efficient than its traditional open hearth counterpart. A natural gas stove can in turn double the efficiency of the improved wood stoves. Thus, with the advent of gas stoves, Third Worlders might be able to double their energy effectively used in cooking, while decreasing their gross energy consumption by a factor of four.

ACQUISITION (IMPLEMENTATION)

Based on the criterion of addressing high demand areas, transportation must be a priority in industrialized nations, especially the United States. Research into new efficiency technologies in this sector is merited to improve efficiency without sacrificing benefits.

Evaluation of the benefits of various forms of energy consumption is also merited. Only casual observation is required to recognize that the average personal vehicle purchased has a fuel efficiency far below (at least 50% below) the most efficient models on the market. This offers a large potential for conservation. Some further investigation suggests that some reasons for these consumer preferences are perceived safety of heavier vehicles and the need or desire for more power. Some consumer preferences spawned by successful advertising seem to be entirely nonrational (e.g., the desire for rugged, four-wheel drive vehicles by bankers and lawyers commuting into Los Angeles). While some preferences may not have an apparent rationality, the reality of those preferences cannot be denied. It is possible for advances in engine materials and construction to improve efficiency in vehicles with significant power and size. Research in these areas might be justified if any evidence is seen that any consumer preferences would be realized for the more efficient vehicles. Another area of conservation is to utilize mass transportation, both for humans and material. Questions must be asked as to why existing mass transportation technologies are not employed to a greater extent, particularly in the United States. The policy ramifications will be explored in greater depth in the final chapter.

Areas of technology that hold promise for capturing conservation gains would also be seen in space heating. Improved insulation is commonly discussed as a potential contributor, which can be used in conjunction with passive solar heating. Fuel choice is significant here, too. The best natural gas furnaces can be much more efficient than solid or liquid equivalents, while the most efficient solid or liquid burning devices can be much more efficient than a standard, older natural gas unit.

In the realm of cooking, natural gas has clear superiority. This is especially true in comparison to open wood fires that prevail in much of the nonindustrialized world but is also true in contrast to electricity. Other domestic applications include improved refrigerators, which can operate at a fraction of the energy input of their predecessors. Since the constant cooling accounts for a large part of the total energy demand of a household, the energy savings can be substantial enough even to see cost savings with a reasonable rate of return for the owner. More efficient fluorescent light bulbs can operate at one-sixth of the energy required for equivalent incandescent bulbs. While lighting is a relatively small demand on a per-household basis, proliferation of the light bulbs could have a noticeable impact on overall regional energy demand. This issue will be considered in greater detail in the chapter on consumption.

Opportunities for conservation can also be very large in the industrial sector. The increased energy prices of the oil embargo period stimulated conservation measures that succeeded in reducing energy requirements in the industrial sector significantly. Cogeneration techniques have been some of the most effective. In general, cogeneration can be viewed as using energy for more than one purpose simultaneously. Frequently, it involves using process heat to boil water for steam to generate electricity. This may involve the heat required for processes such as smelting metals, glazing products, or making cement. If the heat were not captured for cogeneration, it would be released to the atmosphere through a smokestack. A related approach is to utilize a heat exchanger, in which the process heat is transferred to a carrier medium (e.g., water), which can be piped throughout a facility to provide space heat.

Electric generation and transmission have considerable room for efficiency improvement. Less than half of the gross input to a combustion-fueled generation plant is converted to electricity. Major losses are also incurred in transmission. As mentioned previously, superconductor research to improve transmission efficiencies may be one of the more promising areas for the development of new technology.

The facts that huge conservation gains were seen during the 1970s, and that the United States consumes twice the energy per capita as do other affluent, industrialized countries, suggest that considerable potential exists in the conservation resource base.

ENDNOTES

1. Moore, E.S. 1940, *Coal*, Wiley and Sons, NY, pp. 254–256.
2. Ashton, T.S. and Sykes, Joseph 1964, *The Coal Industry of the Eighteenth Century*, Manchester University Press, Manchester, pp. 20,21.
3. Moore, *Coal*, p. 278.
4. *Dictionary of Mining, Mineral, and Related Terms* 1968, United States Bureau of Mines, Washington, D.C., pp. 209, 350, l929, 1240.
5. Moore, *Coal*, p. 281.
6. *Dictionary of Mining, Mineral and Related Terms*, op cit, pp. 181, 1154.
7. Moore, *Coal*, pp. 287–290.
8. Stacks, John F. 1972, *Stripping*, Sierra Club Books, San Francisco, pp. 20, 21.
9. Yergin, Daniel 1991, *The Prize*, Simon and Schuster, NY, numerous citations.
10. *Our Industry, Petroleum*, British Petroleum Corp., p. 96.
11. Giddens, *The Early Days of Oil*, Princeton University Press, p. 6.
12. Moore, Preston, *Drilling Practices Manual*, PennWell Books, Tulsa, OK, pp. 109–112.
13. *Petroleum Data Book* 1990, American Petroleum Institute, Section XXI, Table 14.
14. "Unocal to Close Sole U.S. Commercial Oil Shale Plant", *Oil and Gas Journal*, April 8, 1991.
15. ibid.
16. World Commission on Environment and Development 1985, *Our Common Future*, Oxford University Press, NY, p. 190.

17. OECD 1988, *Environmental Impacts of Renewable Energy Sources*, Paris, p. 47.
18. Poulos, Christine 1992, "A Study and Comparison of Coal and Wood Fuels and their Potential at the University of Rochester," Honors Thesis, University of Rochester.
19. Anderson, Dennis, *Economics of Reafforestation*, pp. 20–24.
20. Steinhart, Carol and John 1974, *Energy: Sources, Use and Role in Human Affairs*, Duxbury Press, North Scituate, MA, p. 96.
    Smil, Vaclav 1984, *Energy, Food, Environment*, Clarendon Press, Oxford, pp. 287–294.
21. El-Gohary, Faitma and Nasr, Fayza 1982, "Bioconversion of Organic Waste," *Alternative Energy Sources IV*, vol. 5, ed. Veziroglu, T. Nejat, Ann Arbor Science Publishers, Ann Arbor, Mich., p. 315.
22. Connet, Paul and Ellen, "If Incineration is the Answer, You Asked the Wrong Question," *The Waste Knot*, 1994.
23. Ramage, Ann 1983, *Energy: A Guidebook*, Oxford University Press, NY, pp. 215, 216.
24. Butti, Ken and Perlin, John 1980, *A Golden Thread*, Cheshire Books, Palo Alto, CA, pp. 19–27.
25. Steinhart, op cit, p. 230.
26. Cook, Earl 1975, *Man, Energy, Society*, W.H. Freeman Co., San Francisco, pp. 63–69.
27. OECDNEA/IAEA 1983, *Uranium Extraction Technology*, Paris, pp. 17–32.
28. ibid, pp. 62–65.

CHAPTER 3

TRANSPORTATION AND STORAGE

Energy must be available when and where it is needed. Energy production sites are often distant from the population (consumption) centers. Inevitably energy is lost or spent in the transport phase. This factor inhibits the development of very remote sources and should be considered carefully in assessing the net efficiencies and impacts of massive sources with low energy intensity, such as biomass.

Transporting combustion fuels from the point at which reserves are extracted to the point of consumption is an important issue. The most conspicuous problem with moving fluid fuel is the potential for major oil spills resulting from accidents to the tanker vessels carrying the fluid. While the solid fuels present less environmental risk in transport, their lower energy content per unit weight mean that more energy is spent in moving these resources.

Energy storage systems can be described by their energy and power densities. Energy density is the quantity of energy per unit mass (e.g., Btu/lb). For example, how heavy would a battery (or a fuel tank) have to be to store enough electrochemical energy to carry a car 500 miles? Power density refers to the rate at which energy can be extracted from storage, in terms of power per unit mass. Power is energy transformed per unit time (such as Btus per hour or kilowatts). To build on the previous example then, how much battery mass would be required to enable the car to accelerate from 0 to 50 miles per hour in 6 seconds?[1]

The nondepletable resources, whose energy flux is tapped

by humanity rather than drawn from reserves, suffer from discrepancies between flux cycles and demand cycles. Indeed, demand is very nonlinear across time, and having energy stored to meet demand is essential. This is a factor that generally favors the combustion sources, as they can be readily stockpiled. The technologies to store heat and electric energy are crucial to the ability of solar or wind resources to take dominant market shares as primary energy providers.

While many of the premier solar applications are for use onsite, large-scale applications do call for conversion to electricity and moving the electricity to the consumers. Electricity is the carrier for most hydropower as well as geothermal energy and wind power. It has no visible emissions (but does experience efficiency losses), is not easy to store, and invisible electromagnetic radiation emissions may produce health hazards.

COMBUSTION FUELS

COAL

Transportation and storage are factors that favor the combustion fuels. The realities of energy use are that most tasks are location and time specific. In fact the advent of the locomotive, the steamship, the automobile, and the airplane made the transportation sector one of the largest consumers of energy and made portability a critically important factor. Even more fundamentally, though, the ability to transport energy resources to intensive consumption sites was key to the industrial revolution. Steel cities grew in the coal-mining regions of the United States, because so much energy was required in steel mills that more coal mass was required to process the steel than the weight of the ore itself. Thus, transportation costs were minimized by locating the mills near the source of coal and shipping iron ore to those sites. As oil and gas took over the energy markets, their ease of transport changed the priorities for steel mill siting.

Since coal exists in a solid state, no special technology is required to store or transport it. Centuries ago, it was carried in carts, or even in sacks on a person's back. Indeed, it could be handled exactly as firewood and charcoal. The success of a new resource in displacing its predecessor is always expedited by matching the predecessor's use and handling characteristics as closely as possible, and coal clearly had that advantage.

As the scale of coal mining increased, tipples began to be used at the mine entrance. Originally, the tipple was simply a huge bin into which mine cars of coal could be tipped, hence the name. The tipple was particularly effective for draft mines which were dug into the side of a hill. In this case, the coal could be dumped into the tipple just under the mine opening, where it could be stored until a cart or wagon was driven under it. Then trap doors at the bottom of the tipple would be opened, allowing the coal to fall into the cart. Gravity did the work of loading. With the advent of railroads, this technique of storage and transfer remained important. Now, the tipple is the storage and transfer facility for coal and is usually located near the mine entrance. Large tipples are found even at strip mines, where coal can be dumped into passing rail cars. Even though this requires lifting the coal into the tipple, it is more expeditious; the tipple doubles as a collection and storage site.

Coal lumps are commonly crushed or broken into relatively uniform size before entering the tipple in modern mining operations. Crushing, which makes the fuel ultimately burn more efficiently, is typically the only processing required before retail. Coal often contains significant portions of sulfur and mineral ash, and it is desirable to remove these contaminants prior to sale and use. Unfortunately, the contaminants are normally found closely mixed with the carbon-bearing constituents such that it is seldom feasible to perform such processing. (Occasionally, coal contains sizable pyrite nodules which account for most of the sulfur and can be separated by mechanical processes.) Although modern applications ultimately use crushed coal, large lumps were preferred on the market into the early part of the 20th century, and miners were only paid for extracted coal that was too large to pass through a sorting screen.

Coal is usually moved in *unit trains* of 100 cars each, with each car carrying approximately 100 tons. The unit trains are dedicated to their coal runs, carrying a load to the consumption point and promptly returning (empty or *dead-head*) to pick up another load. The lines of coal-laden cars, which can seem endless to a motorist waiting at a rail crossing, may transport the product 1,000 miles or more to a buyer. Currently, the richest coal fields in the United States are in the Rocky Mountain states, with modest local consumption due to low population densities. Thus, a great deal of the Rocky Mountain coal is destined for markets in distant, more populous states.

Where waterways exist, barges and freighters offer very efficient alternative forms of transporting the massive quantities of coal. Moving coal was one of the reasons for creating the artificial waterways of the Erie Canal system along the eastern Great Lakes. The ability to move coal (and other massive products) cheaply along the vast, interconnected lake system explains the growth of numerous metropolises in this belt of the United States. Water transport is much more efficient than land transport, although perhaps less direct. (Natural waterways are not laid out for transport convenience and canals are very limited.) In the 1980s, barge transport was second to rail transport of coal in the United States, handling approximately 16 to 17% of U.S. coal.[2]

There are several differences between barges that commonly transport coal on inland waterways and freighters that move products across the seas. Barges are flat-bottomed rather than v-bottomed, and barges are not self-propelled. The barge has the advantage of being a very inexpensive vessel, and can ply very shallow waters. The drawback to the flat-bottom is that it cannot handle wave action, and thus has essentially no oceangoing application and very limited application on the stormy Great Lakes.

Pipelining • In the 1970s, an even more efficient means of moving coal received serious attention—pipelining. Pipelining, effectively the new version of an artificial waterway, is necessarily more efficient than any kind of surface transportation. Generally, this can be claimed because the only mass moved is the product itself rather than massive vehicles. However, coal does require a carrier fluid since it is a solid. Water was proposed as the logical carrier in a plan that came to be known as the *coal slurry pipeline.* The water carrier adds to the mass being moved, offsetting the savings of vehicle weight. Yet the less obvious difference between the two modes of transport is the gross inefficiency of internal combustion engines used to power most surface vehicles. The pumps that move fluid through pipelines operate at much higher efficiencies.

The idea of the slurry pipeline is to pulverize coal to a particle size that could be held in a slurry, or suspension, in water. Because coal is essentially insoluble in water, some surface active agents are required to maintain the suspension. The task of mixing coal with water is aided by the fact that the density of coal is not very different from the density of water. The slurry can then be pumped efficiently through the pipeline to distant consumers. Finally, the slurry must be broken and the coal separated from the water at the point of the end-user.

One insurmountable obstacle the pipeline encountered was objection to the vast water requirements. A major slurry line was intended to run from the coal region of the Rocky Mountains to the eastern United States. Since the Rockies are semi-arid, local residents objected to sending their limited water supplies to the much more humid east. Proponents of the line tried to assure residents that no potable water would be used. Rather, saline water would be drawn from wells to make the slurry. Nevertheless, residents questioned whether subsurface aquifer drainage might affect the groundwater cycle, perhaps drawing shallower fresh waters down into the depressured saltwater zones. Studies could indicate the unlikelihood of this outcome, but not the impossibility. It was not feasible to gather enough information on the subsurface structure and permeability regimes to eliminate the possibility. For instance, if an undetected vertical fracture connected the shallow aquifer with the deep one, it might permit the fresh waters to flow down to recharge the produced saline aquifer. The inability to address this concern adequately was the death knell to the proliferation of slurry pipelines, although the rail companies also opposed the pipelines for economic reasons. Ultimately, the method used to prevent the pipelines was lobbying to deny eminent domain rights to the slurry pipeline. Eminent domain is the law by which highways, railroads, and pipelines can force landowners to lease or sell rights-of-way through their property.

Currently, the Black Mesa line running from Utah to California is the only significant line in operation. It carries about 5 million tons of coal per year, the equivalent of 1.5 unit trains per day.[3] So, to this day, the vast bulk of coal is moved on trains and barges. Storage of the solid fuel is simple and straightforward; it only needs to be piled. Coal is altered enough from its original vegetal state that it is not appealing to termites or their like. However, coal dust can be a problem and can potentially ignite in the storage area.

OIL AND NATURAL GAS

The transport of fluid energy sources, especially over land, can be much more efficient than the transport of solids, yet it can produce far more immediate problems. In the first decade of the oil industry in Pennsylvania, oil was moved and stored in wooden barrels. The 42-gallon barrel that held petroleum in the early days is still used as the unit of measurement for volumes of oil, though no oil ever finds its way into such a barrel anymore. Drake's famous premier oil well produced a modest 10 barrels of oil per day. That doubtless provided good work for the local coopers, but imagine the impact on barrel makers just two years later when the first really large producer gushed in at 3,000 barrels per day! Producers could not begin to buy barrels to contain their production or even to build storage tanks adequate to hold such volumes.

Oil overran the hastily erected containers, pouring down the hills into streams and rivers. Production grew more rapidly than demand, and the containers soon cost more than the black gold they held.[4]

In the latter part of the 19th century, roads were intended for horse and buggy traffic. In cities, they were likely to be cobblestone or brick paved, but in the hills of western Pennsylvania where the oil industry began, dirt roads were a great obstacle to transporting oil to consumers. *Corduroy* roads were constructed by laying logs across the roads to suppress the mud and to support the heavy oil wagons. While these roads minimized the tendency for wagons to become mired in mud, they were not particularly long-lasting.

As oil flowed from the ground, virtually uncontrolled in the early gushers, the shortage of barrels and slowness of moving the product by wagon led to bottlenecks, causing vast amounts of oil to be held at the wellhead. It was common for oil to overflow storage vessels because the gushers were often allowed to flow unchecked until the pressure subsided. Petroleum production before Drake's well existed, but it merely consisted of skimming natural seeps from creek surfaces and the like. The first wells were drilled to reach greater supplies of the product; thus, having to mop it up from the surface was not radically different from the earlier, more natural process. Some producers even dammed ravines to serve as collection and storage basins for the oil runoff. Farmers downstream most likely did not appreciate this technique, and even without environmental regulations, most operators sought to contain their production more cleanly and effectively.

With the shortage of barrels and rapidly rising haulage rates, it did not take long for the first pipeline to be laid. Between 1863 and 1865, numerous field gathering lines were installed and connected to a trunk line which carried the oil on to the rail head. The first large-scale line was the Tidewater, connecting the oil region of western Pennsylvania to an eastbound Pennsylvania and Reading Railroad depot, 109 miles east at Williamsport. Oil began to flow through the line in May, 1879. Previously, the largest diameter pipe used was 3 inches, and no line had extended more than 30 miles or traversed significant topographic relief. The Tidewater not only tripled the distance record but was 6 inches in diameter and crossed the Appalachian Mountains—in mid-winter.[5]

The driving motivation for the record-setting line was not efficiency, environmental preservation, or even cost-savings: it was to break the control being acquired by J.D. Rockefeller's Standard Oil. Once Rockefeller made Standard sufficiently large, he made infamous kick-back and rebate deals with the railroads. These deals not only gave Standard lower rates for transporting oil to their refineries, but gave Standard a portion of the premium paid by its competitors to haul their product.[6] This market control gave Rockefeller's company an effective monopsony—the company bought such a high proportion of the total oil production that it could set the price for crude. The first pipelines had been motivated largely by the grossly inflated haulage fees exacted by the local teamsters. So, under the new market pressures brought to bear by Rockefeller, the producers got together to build the historic Tidewater line to break his control and open new markets. Interestingly, Rockefeller's early Standard Oil was not an oil producer; he believed that the money was to be made in refining and marketing. Unfortunately for the numerous small producing companies, Rockefeller responded quickly to their initiative. Within two years, Standard laid pipelines to Cleveland, Buffalo, Philadelphia, and New York City, restoring Standard's downstream control.[7] Rockefeller's competition-busting exploits subjected marketing competitors and producers alike to cutthroat practices and ultimately led to national trust-busting legislation. Standard's practices broke the backs of many companies whose leadership tried to remain independent and invoked great indignation against unfair practices. However, the practices that sought to choke competition were described by Rockefeller as creating order and efficiency in a chaotic business. Indeed, the struggle for order had a side benefit of eliminating possibly massive environmental damage by implementing pipelines that could move tremendous volumes of oil with much less spillage.

Pipelines quickly dominated intracontinental transport. They proved to be advantageous in nearly every respect. They lost less product, which meant less environmental damage, they could move vast quantities, and they required less energy input and ultimately less materials. As environmental concern grew legislative teeth, engineering design responded by making pipelines even more secure. When a consortium led by ARCO, Exxon, and British Petroleum pooled resources to lay the Alyeska pipeline from the North Slope of Alaska to the nearest open water port of Valdez, pressure-sensitive valve activators were installed at frequent intervals. If a leak occurs, pressure in the line drops and a valve closes automatically upstream of the leak, preventing more oil from flowing to the hole. It thus becomes possible to minimize the amount of petroleum that could spill in any given incident.

Though the security of the Alyeska line has been questioned in the early 1990s by reports of inadequate inspections and corrosion control, the technical possibility of making the line very secure exists. The Alyeska line employed an impressive array of engineering technology 20 years earlier, not only to minimize the risk of significant leakage but to address a broad array of environmental issues. Currently, it is not clear to what extent this famous pipeline is becoming a hazard. If it is, the technology exists to repair it, if the economic will exists.

Obviously, the larger a pipe's diameter, the more fluid it can transport. Also, the more force applied in pumping the liquid or compressing a gaseous fluid, the more can be transported in a given pipeline. (The generalized equations engineers use to design pipelines are shown in the Appendix.) The cost of a line is proportional to the size and strength of the pipe selected. The construction work involved in laying the line is likely to be a large part of the project's cost, so the lines installed should be of adequate capacity to meet anticipated delivery needs. If substantial additional field development is expected, the line should be sized to carry the total production. In the case of lower income countries, production is likely to be demand-limited, so the line should be optimized to meet anticipated energy demands at the delivery point.

Most modern lines are steel, but small-scale applications now sometimes are met with high density polyethylene lines. Polyethylene can withstand much less pressure than steel pipes and is limited to relatively small diameters, but it can be installed for much lower costs. Rather than employing heavy trenching equipment and welders to secure each connection, the polyethylene lines can be installed in a narrow ditch made by a plow-like device, and the connections at each joint of pipe can be made with a simple heat fusion instrument that requires less time and skill than welding. This technology has shown some promise for bringing the benefits of piped gas and oil to the small markets of lower income countries. A modest polyethylene line installed by the Mozambican government oil company (*Empresa Nacional de Hidrocarbonetos*) in 1991 now provides gas to a town of 50,000. The gas is from a field discovered 30 years earlier but previously not deemed cost effective to place on production.

Tankers • While pipelines offer an efficient transcontinental means of moving fluid fuels, increasing volumes of oil have moved between continents since the start of the 20th century. Pipelines crossing the oceans are too costly under current economic and technologic constraints. Thus, shipping is the only viable solution to moving the product from distant producing regions to consumption centers.

The first successful bulk oil tanker was the Zoroaster, built by Ludwig Nobel to move Russian oil across the Caspian sea. Launched in 1878, it was a tremendous improvement over loading wooden barrels of oil or more volatile and explosive kerosene onto ships.[8] During World War II, the United States built 525 tankers to supply the allied forces. Called the T-2, each tanker carried 16,000 tons of crude, fuel, and supplies (deadweight tons). These vessels were close to 400 feet long, and many continued to work well into the 1970s. In 1962, the tanker Manhattan was commissioned at eight times the capacity of a T-2. It carried 116,000 deadweight tons but was still barely half the size of a supertanker. In the 1970s, oil tankers became amazingly large. The *so called* supertanker dominated the market. A supertanker can be more than 400 yards long, hold perhaps more than 2 million barrels of oil, and take more than a mile to stop—at half speed.[9] The industry does not commonly use the supertanker designation, rather referring to Very Large Crude Carriers and even bigger Ultra Large Crude Carriers (VLCCs and ULCCs, respectively.) The increase in tanker size was not only based on the notion that "if big is good, bigger is better," but was also a response to the huge import volumes of the industrialized nations which were matched by the enormous productive capacities of supergiant fields in the Mid-East, Alaska, and south Asia. Ever-increasing international demand seemed to call for ever-larger economies of scale, bigger ships to haul more oil to the industrialized, consumptive nations. Figure 3-1 shows a variety of oil tankers, offering a notion of the relative size and capacity of the supertankers.

The impacts of supertanker accidents can be staggering. Several notable accidents have occurred but none received more global press (led by the American news media) than the first supertanker disaster in American waters: the infamous Exxon Valdez outside of Valdez Harbor, Alaska. Although the Valdez was not among the largest of supertankers (barely qualifying as a VLCC) and barely spilled one-sixth as much as the Amoco Cadiz 11 years earlier off the coast of France, the accident raised a great deal of furor about the shipping of oil.[10]

Ironically, much of the hostility about the transportation accident was directed at the unrelated practice of offshore drilling. Hostility can perhaps more reasonably be directed at the use of supertankers. However, the simple logistical reality is that, with given import demands, opting to use smaller tankers mean more tankers, more traffic, and almost certainly more accidents—though supertankers mean potentially larger accidents. Any nation that

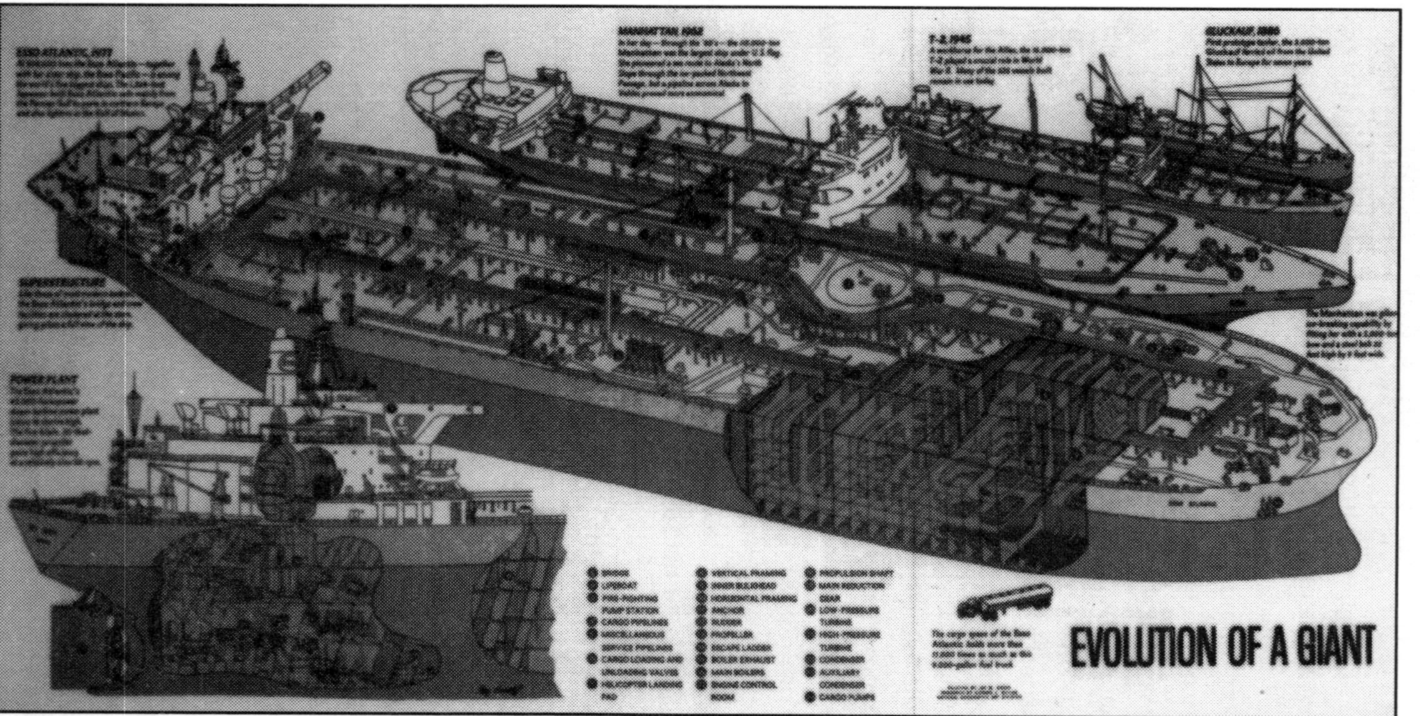

Figure 3–1 A chronology of oil tankers.
The early tankers were very small. By the early 1970s, some of the oil tankers were among the largest ships afloat. For comparison, a modern aircraft carrier is about 1,000 feet long.
Courtesy *National Geographic*.

consumes more energy than it produces must import to provide for the deficit, and moving oil across the oceans is a high-risk activity. The debate about the policy choices for environmental preservation will be saved for the impacts chapter.

Modern supertankers incorporate a number of design features to minimize the likelihood and severity of accidents. These ships have fully separated multiple holds (Fig. 3.1), so that damage to one part of the ship does not necessarily endanger the entire cargo (hence, the Valdez spilled barely one-fifth of her cargo on that tragic morning). The ships are built incorporating computers and navigational, safety, and monitoring equipment. Earlier vessels lacked sophisticated equipment, and it really took years before the development of the first truly successful tanker design. The first efforts to ship oil across the Atlantic consisted of loading wooden barrels full of oil, or even more likely refined kerosene, into the holds of conventional cargo ships. The barrels could shift and rupture, spilling their contents into the remainder of the hold where open lanterns could ignite the fumes from kerosene. When the concept of sealed hold tankers was developed, some of the prototypes did not have any sort of baffle system. Heavy wave action could create wave action in the liquid cargo. The waves in the hold could add to the momentum generated on the ship by the ocean waves and increase the likelihood of capsizing. Modern technology has addressed these problems.

Natural Gas Transportation • Natural gas is moved very easily and efficiently through pipelines. Before the advent of the petroleum industry, coal gas was piped into larger cities for street lamps and some household use. Gas, though, was not a pre-eminent energy source. In the early days of the oil industry, immense quantities of gas were flared, and that practice has yet to be entirely displaced. Even when a large proportion of American homes switched to natural gas for cooking and space heating, gas remained a secondary commodity to oil. The suppression of the gas market was certainly aided by the government's decision in

1954 to control natural gas prices. Other impediments, though, were problems with storing gas and with transporting it beyond the range of pipelines.

The first commercial use of gas was a manufactured gas made from coal (as described in the previous chapter). The Scottish Engineer William Murdock is commonly viewed as one of the leading pioneers in the use of gas, installing gas lights in his home in Cornwall, England in 1792. Many others experimented with the use of gas at essentially the same time, even with some earlier success, such as George Dixon's lighting installed 145 years earlier. It was Murdock, though, whose success seems to have provided the greatest impetus to the creation of a gas industry. In those early days, wooden pipelines were constructed to carry the fluid. In the first two decades of the new industry, success was limited by an inability to store the product—excess production was simply flared. In 1816, Samuel Clegg developed the first gas storage containers. They depended on water seals which contaminated the gas but were not replaced by dry storage containers until the turn of the 20th century. Steel pipelines did not take over the pipelining market until the development of the Mannesman process to make seamless pipes. While piping gas was prevalent in England nearly 200 years ago, bamboo pipes are said to have carried gas in China as much as 3,000 years ago.[11]

The early wooden pipelines and the steel lines that succeeded them were generally small and operated at low pressure. The gas often simply flowed through the local pipelines with some form of regulator controlling the natural pressure from the well. Even at the writing of this book, it is not uncommon for the productive life of a well to be determined by the point at which reservoir pressure has declined to the point that it can no longer push gas to the surface at a pressure exceeding the pipeline pressure. In the case of large fields, it can be cost-effective to install a compressor that takes produced gas at lower pressures and brings it up to line pressure.

As demand for gas grew, it became cost-effective to lay large-diameter and high-pressure trunk lines. There are many gas pipelines in existence with diameters greater than three feet, and some exceed five feet in diameter. Gas flowing in large volumes through these pipelines to feed the massive industrialized world's markets has its pressure boosted at compressor stations along its route.

In order to store natural gas in significant quantities, or to transport it beyond pipelines, gas must be compressed under tremendous pressure or liquefied in extreme cold. Liquefying natural gas (LNG) began as a storage measure but has proven to be a viable means of transporting natural gas as well. LNG tankers bring Japan much of its natural gas from fields around the world. The typical oceangoing LNG tanker commonly has dimensions of about 125,000 cubic meters, which holds about three-quarters of a billion standard (atmospheric conditions) cubic meters of gas, or nearly 8 billion standard cubic feet.[12]

Compressed natural gas (CNG) is receiving attention as a potential vehicular fuel, because, like gasoline, CNG permits storing a sufficient amount of energy in the limited space of an automotive fuel tank. Compression at safe pressures is not as efficient a means of storage as liquefying the gas, but the supercooled conditions required for LNG are very difficult to maintain on the road. So, despite the lesser efficiency, it is still possible to put enough CNG in a tank to have reasonable cruising range. Both CNG and LNG are being tested in bus fleets. City bus fleets are not faced with having to find filling stations that carry their fuel. The analyses seem to be indicating good performance of the natural gas fuels but currently at slightly higher cost than diesel.[13]

BIOMASS

Since most biomass is used for energy in the forms of firewood or charcoal, transportation is normally by truck, cart, or hand. Firewood is readily stored by stacking, and in America and Western Europe, wood is traditionally stacked in cords. A cord is 4 feet high, 4 feet deep and 8 feet wide. A typical cord of pine has a chemical energy content of approximately 27 million Btus. Wood can go through numerous cycles of taking on moisture and drying, though it must ultimately be dry to start burning readily. However, the main problem with wood storage is one of protecting it from small creatures, especially termites and tunneling ants.

Since firewood has such low-energy density, a great deal must be acquired and transported; the fuel consumed by trucks tends to offset the energy gained from the firewood. On a typical firewood-gathering expedition of 20 miles in each direction, the fuel consumed by the truck only amounts to 10% of the chemical energy content of the wood; it could be much more or somewhat less, depending on the condition and efficiency of the truck. In countries where people depend on firewood, the mass of the cargo sufficiently impinges on firewood sellers' economics that wood is often converted to charcoal prior to shipment from the forests.

Charcoal is much lighter than raw wood, and since much of the mass driven off in the conversion process is water, charcoal has a higher energy density (more Btu/lb). A consumer preference for charcoal probably adds incentive to the conversion decision, but probably not as much as the transport efficiency. (When consumers pay upwards of one-third of their income for cooking fuel, they are unlikely to pay any noticeable premium for the convenience of charcoal.) It is possible that the transport efficiency gains in hauling less dense charcoal may be sufficient to offset the energy losses in the inefficient conversion process. The efficiency gains depend on the distance of the haul. In lower income countries with diminishing forests, the fuel may be hauled as many as 100 miles. People (most frequently women) walk as many as 20 kilometers (12–13 miles) in each direction to gather firewood for the family meals. In some firewood deficient regions, this activity has been estimated to take as much as 30–40% of women's time. In the sense of energy conversion, human effort is much more efficient than vehicular transport, but productivity suffers as more human time is spent in gathering firewood (Fig. 3-2).

The future of biomass fuels doubtless lies in conversion to secondary fluid fuels (alcohols or gas). Once biomass is converted to a secondary fluid fuel, it shares the same transportation advan-

Figure 3–2 Women carrying firewood in Kenya.
Courtesy Mrs. Nancy Polling, Rochester, NY.

tages and disadvantages of oil and gas. Gathering and transporting the solid biomass in order to get it to the conversion facility is likely to present significant limitations to the applications. Fecal material is an excellent biogas feedstock, but the manure produced by free-range animals is too dispersed for viable collection in most applications. Generally, biogas is most viable when consumed at the point of production from an existing concentrated waste stream. Alcohols produced from plant matter are likely to be destined for mixing with petroleum products for use in internal combustion engines and will be transported in the same manners as oil.

NONCOMBUSTION SOURCES

SOLAR POWER

Solar energy is found in the electromagnetic radiation from the sun which arrives at the earth's surface concentrated in the narrow visible range of the spectrum (Fig. 3–3). Since existing technology (and scientific understanding) sees electromagnetic radiation as energy in flux, it clearly cannot be stored without converting it

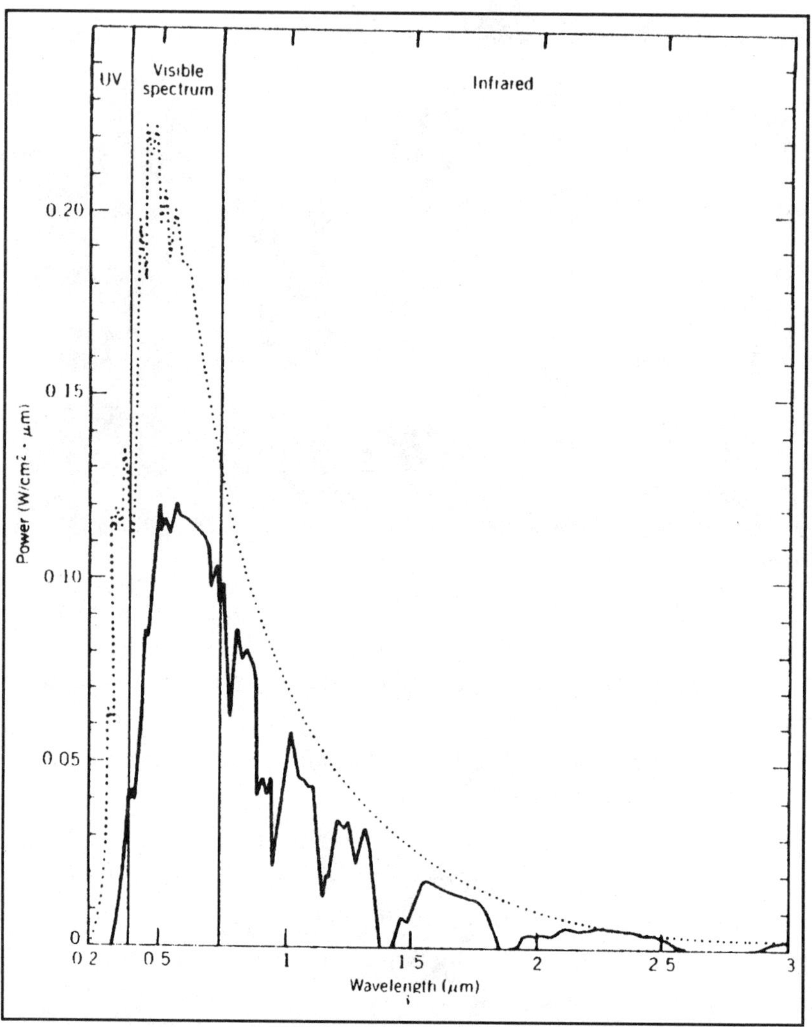

Figure 3–3 The electro-magnetic spectrum.
Solar radiation reaching the earth's surface is concentrated in the visible portion of the spectrum, but much of that which is reemitted by the earth's surface is in the infrared (heat) portion of the spectrum.
Courtesy Jack Kraushaar and Robert Ristinen, *Energy and Problems of a Technical Society*, p. 149.

to another form. Photons, which can be thought of as energy packets, can interact with matter in a number of ways. They can strike nuclei, causing them to vibrate more rapidly, which is observed as heating of mass. Effectively, the heating of air masses in this manner

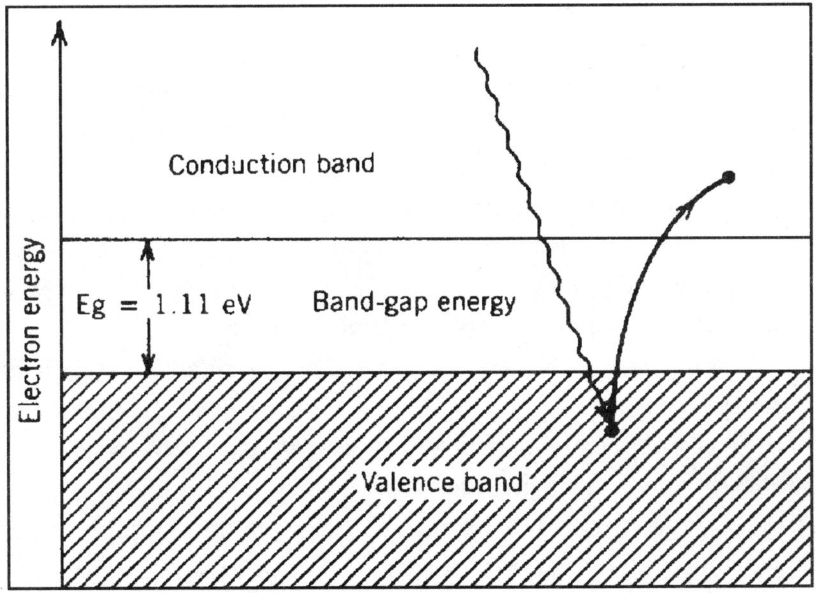

Figure 3–4 Energy bands of electrons.
When photons impart just the right amount of energy to a valence electron, it can move up into the conductance band. The electron flow creates an electric current. Courtesy Jack Kraushaar and Robert Ristinen, *Energy and Problems of a Technical Society*, p. 176.

is what causes most wind. The photons can also interact with electrons, imparting energy to the electron. In the case of silicon and other semiconductors, the outer electron valence is essentially full, and the electrons are not prone to flow. However, if an electron receives more than 1.1 eV (electron volt) from a photon, it can jump out of the valence orbital into the conductive band, permitting electric current flow. Figure 3-4 illustrates how electrons can be activated to flow by solar energy. Photovoltaic systems use this fact to convert incoming solar radiation promptly to electricity.[14]

Storage • The interaction of solar radiation with chlorophyll and other pigments in plants yields a transformation of the energy in flux to chemical energy, which is stored in the tissues of the plants conducting the photosynthesis and in the animals that eat the plants. In general, the light passing through an electron cloud imparts enough energy to the electrons to mobilize them sufficiently to bond in the formation of high energy molecules. This is a very efficient storage process which produces not only all of the biomass energy on earth but, through the eons, has stored all of the fossil fuel energy as well. The natural process is efficient enough that artificial photosynthesis is not likely to become competitive, but the advantages of storage in the form of chemical energy are compelling; chemical energy can be transported with potentially no loss and can be stored indefinitely.

Even passive solar space heating is best designed to store heat energy during the day to be released into the area at night. The most basic applications involve placing a massive, dark wall opposite the window that provides solar gain. Since an object appears dark because it does not reflect light in the visible portion of the spectrum (and a majority of solar radiation occurs in this portion), a dark object absorbs a great deal of solar energy. The energy is stored in the enthalpy of the material. In the case of passive solar heating, this storage is often provided simply by a block or stone wall painted black. When the temperature of the surroundings drops below the temperature of the storage mass, it radiates heat to the area, maintaining a more constant temperature.

A variation of this technique is the *Trombe* wall (named for its developer, Dr. Felix Trombe). A massive, dark wall is just a short distance behind a glass wall with southern exposure. The massive wall has vents which permit convective air flow. When ambient temperatures are cool, vents at the top and bottom of the wall are open, and warm air circulates through the top vent into the living space while cool area flows through the bottom vent back into the inter-wall space to be heated by the sun. After sundown, the vents can be closed to allow the stored heat in the wall to radiate to the living space. (Inevitably, heat will also radiate to the space on the south and some will escape through the glass.) One advantage of the Trombe system is seen in the summer, when the top vent in the wall can be closed, while the bottom vent is opened, as well as a vent at the top of the glass wall. The heated air in the interwall space rises to escape through the glass vent. It creates a *chimney* effect, drawing cool air into the living space through a northern window to replace air being drawn from the house through the bottom wall vent. Tests have shown that a house with a direct southern exposure and a Trombe wall can achieve 100% solar heating during the winter, though problems have been observed with heat buildup in the summer, requiring additional venting.[15] The Trombe wall system does not need to be extremely expensive if the application is well-suited. In new construction, it involves little more than building the southern exposure house wall of masonry and attaching a greenhouse in front of it. Aesthetic issues seem to be the significant limiting factors. The Trombe house may present a rather flat, stark southern face.

An even more efficient thermal storage mechanism utilizes a phase change. When a solid melts or a liquid evaporates, a great deal of energy is required. Conversely, a great deal of energy is released as a vapor condenses to its lower energy liquid state or a liquid freezes to its even lower energy solid state. Thus, if the dark massive object contains a solid with a melting point near room temperature, the solid tends to melt under the absorption of solar radiation. When it refreezes (at room temperature), it releases the energy stored in the phase change to the room. *Glauber's Salt* (sodium sulfate decahydrate) is one of the most effective of these materials. It has a melting point of about 93°F and a heat of fusion (the amount of heat required to convert solid to liquid or released in the opposite conversion) of 100 Btu/pound. This greatly exceeds the thermal capacity of water on a unit weight (or volume) basis.[16] The simplest applications of this technology involve placing black chemical drums filled with Glauber's Salt in a room, positioned to receive a great deal of winter sun. The drums absorb the sunlight as heat, and the salt crystals store it and reradiate it to the room when the temperature begins to fall. Again, the aesthetics of black drums in the living room can be questionable, but the means can be found to blend the vessels into a reasonable decor; one option is to employ dark, hollow columns filled with the selected material. The remaining problem is the tendency for Glauber's Salt and other phase-change materials to degrade and fail to re-crystallize with time.

Transport • Solar heat can be used to generate steam (or the vapor phase of more volatile liquids) to drive a turbine and convert energy to electricity. Such processes are called solar thermal electric conversion (STEC) processes. Electricity is the carrier energy form produced both in STEC and in photovoltaic applications. An advantage of the STEC processes is, that since they employ a turbine, the electricity can be directly produced as AC (alternating current). AC can be readily transmitted several miles, while the direct current produced by photovoltaic cells is not conducive to long distance transmission.

Heat absorbed by the molecules of a carrier fluid (e.g., water or air) can be transported short distances. As has been discussed in detail regarding geothermal sources, diffusion of heat in transit is inevitable. Vibrating high temperature molecules contact the molecules of the pipe or other containment material, imparting a portion of their heat to that material which imparts some of its newly gained heat to its surroundings, conducting heat away from the heat-carrying fluid. Therefore, when heat is the desired energy form, it can only be moved effectively across very short distances.

Solar water heating is one viable application of gathering solar energy as heat and moving it a short distance to storage, to be moved another short distance to the hot water taps on demand. To accomplish this, most solar water heating systems employ an electrically driven pump. The pump pushes water through the elevated solar collection panel and into the hot water storage vessel/heater. The pump consumes much less energy than conventional batch water heaters, but the electricity required must be charged against the gains from this type of solar application.

In recognition of this efficiency cost, some researchers have designed systems that do not require a pump to circulate the water. Rather, the tendency for water to expand (very slightly) when heated drives the circulation. The process, known as *thermosyphoning,* can keep water circulating, without a pump, if the storage tank is placed above the heat-gathering tubing coils. These systems are passive because of their self-contained energy cycles.

WIND POWER

Wind, the kinetic energy of moving air, also cannot be stored or transported in its original form. In the case of wind, the direct application of kinetic energy to tasks at the point of collection has been used for centuries and has considerable potential for modern use. If the energy is to be transported or stored, it is normally converted to the versatile carrier, electricity. It is then electricity which is either stored or transported.

Most rural wind pump systems include energy storage. They pump water up into a storage tank, adding kinetic energy which is stored as potential energy until the water is withdrawn. Then the stored energy is manifested as pressure at the spigot. It would be possible to place a turbine at the bottom of the storage tank outlet, retrieving kinetic energy from the water. Some consideration is being given to making use of this hydropower storage phenomenon as the storage mechanism for wind-generated power.

HYDROPOWER

The kinetic energy of water, like wind, can be applied directly to tasks as water mills do. Because the movement of water constitutes the conversion of potential energy to kinetic energy, it is possible to contain the falling liquid behind a dam; this encompasses storage as well as concentration of the water's potential energy. The storage process is nearly as perfectly efficient as the laws of thermodynamics allow. However, water storage requires a large volume, and the surface area required to make that volume has been the source of some environmental controversy, especially for large environmentally disruptive dams. Large dams have economies of scale related to storage capacity and power intensity; they produce power cheaply because the costs are spread out across a tremendous amount of energy produced. Clearly, a large dam restrains a large volume of water and stores more potential energy. Similarly, the higher the dam and the deeper the reservoir, the more rapidly energy can be extracted.

It is hypothetically possible to transport the energy rather efficiently to a consumption site by diverting a river. However, diverting any sizable river is a monstrous engineering task, with a large environmental impact. So, once again, moving the energy to the consumer involves conversion to electricity.

Tidal sources are normally tapped in the same general manner as conventional river power, but wave power does not have the same potential to be stored behind a dam. Neither of the marine water applications (tide and wave power) has the same need for storage. Wave action is continuous, although it can increase and decrease in intensity with time. Tidal energy is regularly intermittent, which does require short-term storage to eliminate the periodicity.

Hydropower can also be used as a secondary energy carrier for storage. The surplus power of an intermittent energy source (e.g., solar or wind power) can be used to pump water up into a reservoir or elevated tank, effectively storing the surplus as potential energy. Similarly, the surplus output of a source produced during off-peak times can be stored for later use. During peak demands or times when the primary energy source is unavailable, the water can be permitted to flow back down through a turbine, generating electricity when it is needed. The net efficiencies of such systems can be very high, perhaps as high as 64%. For commercial-scale applications, it is necessary to find or create sites with two large reservoirs at significant elevation differences. The largest existing such facility is in Luddington, Michigan. An artificial lake was created 250 feet above Lake Michigan. During off-peak hours, water is pumped from Lake Michigan into the artificial reservoir, and it reenters the lake through a turbine when additional capacity is required. It is capable of producing 2 gigawatts of energy, with a maximum storage capacity of 15 gigawatt hours.

Efforts are also underway in several places to generate underground storage reservoirs to serve the same purpose.[16] Cost is a serious constraint where topography does not offer easy potential.

GEOTHERMAL POWER

As its name indicates, geothermal energy is found in the form of heat, which is very difficult to transport or store efficiently. Indeed, here lies the major constraint on this energy form. Heat naturally flows from warmer to cooler points; thus, heat is unavoidably given up by the transport medium to its surroundings. Insulation is merely material through which heat flows slowly. Thermal insulation is rather like putting straw in the bottom of a leaky bucket—it slows the loss but cannot eliminate it, so you still have to move quickly with the patched bucket. Even within the wellbore, substantial heat is lost from the produced steam or water to the wellbore surroundings in the early days of a well's productive life. Before long, though, heat accumulates in the rock surrounding the wellbore, and the heat loss slows greatly. Even more heat is lost in the surface lines carrying the hot fluid to the electric generators, and this loss is not ameliorated with time. The air surrounding the steam lines circulates, carrying heat away rapidly and maintaining a high temperature difference to drive the heat loss.

As mentioned in the preceding chapter, thermal expansion of the pipes is an issue of concern as well as thermal losses. Just as the wellhead can grow out of the ground as the geothermal steam heats the steel, the production line carrying steam from the well to the power plant tends to expand. If the production line is straight and is pinned at both ends, it will buckle. Consequently, the surface lines must be built to allow for expansion, either with expansion loops or *grasshopper legs*. Expansion loops are simply joints of pipe preshaped in loops, like those on a roller coaster. When thermal expansion occurs, the pipe at both ends of the loop tend to push together, making the loop larger. Grasshopper legs are pipe joints which turn 90 degrees at each end to a union connection. One side of the union connection has a female-threaded collar (threads are on the inside) fitted over an enlarged end of the pipe, while the other side has an enlarged male fitting (threads on the outside). The enlarged, rounded, convex end of the female side fits into the concave opening in the male side, while the female collar is screwed over the male threads, pulling the convex and concave ends tightly together. This type of connection provides a good seal, while still permitting rotation of one pipe relative to the other. In the case of the grasshopper joints, since the ends are bent 90 degrees, the rotation allows the two pieces of pipe to move in scissor fashion. A pair of such pipe connections together makes the grasshopper leg joint. As the pipes expand, the leg simply bends itself together. Clearly, for either kind of expansion joint, the body of the pipeline must be able to slide back and forth as it expands and contracts. The specially designed pipelines are also insulated to minimize heat loss to the atmosphere. All of these design features add greatly to the cost of the pipeline but can only mitigate inevitable energy losses in transport. Therefore, the steam is never carried far from the wells.

Converting the heat to electricity as near the well as possible is the objective. Then it is transported to the consumer in the secondary form of electricity, as with any of the noncombustion resources.

NUCLEAR POWER

As with combustion resources, fissile material can be stored and transported by simple mechanical means. Losses occur in storage through natural radioactive decay. The same instability of the nuclei which makes a substance fissile causes radioactive decay towards a stable nuclear mass and configuration. The decay is exponential in nature and is measured in terms of the half-life—the time required for one-half of an isotope to decay to its next level. The half-life of uranium 235, the fissile isotope of this naturally radioactive element, is more than 700,000 years. Consequently, the amount lost in storage is certainly negligible, about one ten thousandth of one per cent per year stored.

Efficiency calls for processing the ore at or near the mining site, since only a small fraction of the pure uranium is composed of the fissile isotope. A tremendous amount of energy would be wasted transporting the unpurified uranium. Thus, the ore is purified and enriched at the mine as described in the preceding chapter. The rich, *hot* product is loaded carefully into lead-lined containers, each holding well under the critical mass of uranium to ensure against a spontaneous chain reaction.

Since one year's fuel supply for a large gigawatt nuclear fission reactor is only 5,000 tons, the total transportation cost is much less than for standard chemical combustion fuels. For instance, a comparable gigawatt coal-fired power plant requires more than 2 million tons of coal per year, enough to fill 200 unit trains—about 4 unit trains making deliveries to this power plant every week. (This assumes that the plant operates at an average 70% of its gigawatt capacity.)[17]

Environmental and health issues are central to every stage of nuclear power utilization. Popular perception tends to be that radioactivity is extraordinarily lethal; thus, extraordinary precautions are taken to prevent leakage of stored or transported radioactive material. The handling of radioactive materials must include containment through which the radiation cannot penetrate, as the handling of corrosive chemicals demands special containment. In fact, high levels of radiation are very dangerous and should be handled with care similar to the care with which strong chemical toxins are handled.

Nuclear power is used in peace strictly to produce electricity.

TRANSMISSION OF ELECTRICITY

Electricity is the preferred carrier energy form for most consumers. The preference is related to the versatility and end-use cleanliness of this energy form. Electricity can readily be converted back into heat (from which it most likely came) by passing the electric current through a filament of high resistivity. Any conductor has some resistivity, which is a tendency for the movement of electrons to produce emissions in the electromagnetic radiation spectrum. The emissions from a tungsten filament in an incandescent light bulb in the visible spectrum and the emissions in a toaster in the infrared (heat) range are the desired products, and materials are designed to maximize the interactions that produce these emissions. The same resistivity phenomena are at work in the power lines that transmit the electricity from the point of generation to the consumption point.

In the early days of electric power generation, electricity was generated and used as direct current, in which the electrons flow through a conductor from an electron source electrode towards the other pole of the circuit. Alternating current (in which electrons reverse their flow direction in the conductor many times per second) was found to suffer less resistance loss. In spite of vehement protests by some and fears that it would be incredibly dangerous, alternating current was chosen by electric utility companies.

The transmission of electricity has been a limiting factor since its introduction. Before 1920, transmission losses tightly restricted the area that could be effectively served by any power generating facility. In 1920, however, the advent of high voltage transmission systems helped to mitigate this problem.

The higher the voltage in a transmission line, the less current (amperage) is required to carry a given amount of power through a wire of fixed size and composition. The resistance losses are minimized by forcing more current through the line at a higher rate. This also permits large-scale transmission of electricity through wires of manageable diameter. At lower voltages, the copper or aluminum conductor requirements would dominate the cost of an electrification project. The early high voltage transmission lines operated at approximately 110 KV (110 kilovolts), while modern lines often carry in excess of a half million volts. The problem with such high voltage systems (beyond the obvious hazard to any who might contact exposed lines) is the *corona* effect. At these voltage levels, air near the line becomes ionized. In recent years, the ionization has raised health questions. Even at such extreme voltages though, noticeable losses are experienced with long distance transport of the energy.[18]

A developing technology which seems promising is the use of superconductors. At temperatures approaching absolute zero (the temperature so low that molecules stop vibrating altogether and hypothetically collapse into themselves), some materials lose nearly all resistance to electrical conduction. These are superconductors. While they seem very attractive because of their nearly infinite conductivity, superconductors are clearly not a perfect solution. A significant amount of energy is required to maintain such low temperatures, even as cryogenic technology (the science of cooling things) improves. However, technical obstacles are often not the limiting factor in the development of a technology. A more important criterion is whether the technology offers a net improvement over existing techniques. In this case, the question is whether cryogenic technology can reach the point that less energy is consumed in cooling an extensive transmission superconductor system than is lost in high voltage transmission. The practical corollaries to this question are whether supercooled superconductors can be cost-competitive with high voltage technologies, and what externalities (e.g., health and environmental hazards) must be assigned to each.

STORAGE

ELECTROCHEMICAL STORAGE

Storage is particularly an issue with intermittent primary energy sources, such as wind and solar. Since electricity is the preferred energy carrier, batteries offer one mechanism to convert electricity to chemical energy that can be withdrawn on demand. The term battery properly refers to an arrangement of chemical storage cells. Each cell has an anode and cathode, with some electrolyte between. In the common lead-acid storage battery, an electric current drives positively charged hydrogen ions towards the cathode (negative terminal) of the cell. Sulfuric acid serves as the electrolytic medium in which hydrogen ions are readily dissociated. When current is applied to the cell, the positive ions are driven across a semipermeable membrane to accumulate at the negative pole. In the lead-acid battery, the cathode is elemental lead, while the anode is lead oxide. When the battery discharges current back to the system, positive hydrogen ions (cations) flow back across the membrane towards the lead oxide rod. (Electrical convention refers to a positive current flowing from the cathode to the anode, but physical chemistry shows electrons flowing through the wire from the anode to the cathode.)

The reaction is not perfectly reversible. So, each time the battery goes through a charging and discharging cycle, some of the lead and lead oxide in both poles are converted to lead sulfate during discharge. While charging tends to convert the electrodes back to their original state, some remain unconverted and may fall away in flakes or platelets, leaving the battery unable to take a full charge. Eventually, the battery must be replaced. To the extent that lead-acid batteries are not recycled, they pose a disposal problem, due more to the heavy metal lead than the acid. More advanced batteries use other substances but similar principles.

More than charging cycles, though, energy and power density requirements motivate the quest for new battery types. The energy density in electrical terms is represented by watt-hours/ kilogram. Physically, this represents the fuel tank equivalency of a battery: how far a car could drive per kilogram of battery storage. The power density, given in watts/kilogram (which would be the energy density deliverable per unit time) represents acceleration. The requisite operating temperatures are very high for some batteries that are capable of storing a great deal of energy and of high energy output rates.[19] Filling up a motor vehicle with batteries to provide storage for adequate cruising range and insulation for the operation of the most efficient batteries is a problem. However, a power plant site may easily provide space for batteries and insulation. So, storage may be a factor that favors solar and wind penetration of electric power markets more readily than automotive markets.

Since the vehicular sector dominates American consumption, and because the potentially cleanest and most sustainable energy sources, solar and wind, are intermittent and inconsistent, a variety of storage methods applicable to automobiles do receive attention. The power density requirements of modern vehicles can be easily met by hydrogen storage. As will be described in detail in the next chapter, electricity can be used to perform hydrolysis on water, breaking the water molecules apart into hydrogen and oxygen. The hydrogen can be gathered and stored for later oxidation. This process is attractive because the product of burning (or of slowly oxidizing) hydrogen is simply water. Hydrogen is the lightest, thus least dense, of all gasses at atmospheric conditions; therefore, hydrogen must be stored at very low temperatures or very high pressure to have a significant quantity in a given volume. This can be a problem for mobile applications of hydrogen, as it is for battery storage.

Cryogenic storage of hydrogen at temperatures as low as -253°C is feasible and cost-effective in stationary applications, such as power plants. Unfortunately, the thermodynamic efficiency of conversion back to electricity is a relatively low 40%, as it is for other thermal electric generation processes. When this loss is combined with the efficiency losses in the electrolysis process (about 30% loss) and the minimal losses in maintaining cryogenic conditions, the overall efficiency is not likely to exceed 25% (and that does not include the inefficiencies in generating the electricity to begin with). One means to compensate for the efficiency problem is to regenerate electricity with a slow oxidation process rather than fast oxidation (burning). The slow oxidation of hydrogen can be accomplished in a fuel cell, which employs a catalyst in the reaction, operating much like a battery. The slow oxidation reaction in the fuel cell can be much more efficient in generating electricity, achieving efficiencies as high as 85% in laboratory experiments.[20] Problems remain in making the fuel cells practical for market penetration. Considerable research on fuel cells is ongoing to make them more cost-effective, dependable, and efficient.

KINETIC ENERGY STORAGE

Energy can also be stored as momentum. Kinetic energy can be imparted from one system to another, as a billiard player imparts kinetic energy from the cue to the cue ball; once the cue ball is set in motion, it has kinetic energy that exhibits momentum. Momentum is a function of mass and the square of velocity as is force or energy. The linear momentum of a cue ball is worthless as an energy storage unit, but rotational momentum is a different story. A flywheel rotating on a vertical axis, like a toy gyroscope, can be an effective energy storage device. If it is sufficiently massive and rotates sufficiently fast, it can store a great deal of energy. Since the energy stored is a function of the square of velocity, high speed rotation is important, and the strength of the wheel's composition is more important than its mass. (When mass rotates at high speed, centrifugal forces tend to pull the wheel apart.) Fused silica is a material with extraordinarily high strength to weight ratio (20 times greater than steel).

Flywheels with materials allowing great rotational velocities are capable of storing energy with a very good density—as much as 40 times greater than the energy density of the common lead-acid battery. These systems can also produce very high power densities, which means that they could have vehicular applications. In fact, it is possible to build a flywheel drivetrain system with total mass comparable to the mass of the engine, fuel and drivetrain of a gasoline-powered vehicle, with similar acceleration and range. One advantage of the flywheel vehicle is its ability to recharge itself to some extent while driving. By tying the braking system into the flywheel, the brakes, which usually convert kinetic motion to heat through the friction of brake pads and drums, would convert the vehicle's linear motion back to rotational motion.[21]

The efficiency of a flywheel storage system requires minimizing ambient resistance to rotation. One significant source of resistance is air friction. Flywheel systems operate much more efficiently in a vacuum. Minimizing the friction at the contact points at the ends of the flywheel axis is essential as well. The state-of-the-art in creating frictionless environments has led to flywheels that can store momentum for as much as six months.

COMPRESSED AIR

Energy can be applied to the work of compressing air in a storage vessel. When that air is released, it reexpands, releasing energy which can be imparted to a turbine or piston or some other mechanical action. CO_2 cartridges that power devices such as pellet guns or compressed air powering tools are common examples of this utility. However, this method must be done on a vastly larger scale to be considered a viable energy resource.

In Germany, one commercial-scale use of a compressed air energy storage system is in place. A void in a large underground salt body serves as the storage vessel which contains a volume of approximately 300,000 cubic meters at pressures up to 1,000 pounds per square inch. Since this pressure is approximately 70 times atmospheric pressure, the cavern holds a volume of air that would occupy 2.1 million cubic meters at the surface (ignoring the change in temperature of the compressed gas). The system is capable of delivering 300 MW for 2 hours.

Unfortunately, air compression has significant efficiency losses. One of the most significant relates to the fact that temperature changes due to compression cannot be ignored. The ideal gas laws indicate that PV = nRT, where P is pressure, V is volume, n is the number of moles of gas, R is a unit conversion factor, and T is the absolute temperature. From this formula, it is clear that increasing the pressure of a gas tends to decrease its volume and increase its temperature. In the German application, if the compressed air were allowed to heat naturally, it would threaten the structural integrity of the salt storage cavern. Therefore, energy must be used to cool the gas as it is compressed. Then, the gas must be heated as it is drawn back out of the well before it enters the turbine to generate electricity. If the waste heat generated in the cooling cycle were stored for the reheating cycle, it could significantly improve the efficiency of compressed air storage systems.[22]

In addition to efficiency problems, difficulties with maintaining a high pressure storage system that does not leak limit the applications of compressed air.

ENDNOTES

1. Kraushaar, Jack and Ristinen, Robert 1993, *Energy and Problems of a Technical Society*, Wiley and Sons, Inc., p. 221.
2. International Energy Agency 1985, *Moving Coal*, OECD, Paris, p. 59.
3. Cassedy, Edward S. and Grossman, Peter Z. 1990, *Introduction to Energy*, Cambridge University Press, NY, p. 115.
4. Yergin, Daniel, 1991 *The Prize*, Simon & Schuster, p. 28.
5. Giddens, Paul 1948, *Early Days of Oil*, Princeton University Press, Princeton, NJ, p. 115.
6. Yergin, *The Prize*, pp. 39–47.
7. ibid, pp. 43, 44.
8. ibid, p. 8.
9. Grove, Noel 7/1978, "Giants That Move the World's Oil," *National Geographic Society*, Washington, D.C.; Yergin, *The Prize*, p. 59.
10. *Collier's Year Book* 1990, MacMillan Educational Corp., NY, pp. 110–111.
11. Peebles, Malcolm W. H. 1980, *Evolution of the Gas Industry*, New York University Press, NY, pp. 5–16..
12. Edmonds, Jae and Reilly, John 1985, *Global Energy: Assessing the Future*, Oxford University Press, NY, p. 133.
13. Regional Transit System, Report on Status of CNG Bus Fleet, Rochester, New York, 1994
14. Kraushaar and Ristinen, *Energy and Problems*, pp. 175–177.
15. Tasdemiroglu, E. et al. 1982, "Simulation of Selected Design Parameters of Trombe-Thermal Storage Wall Passive Systems," *Alternative Energy Sources*, vol. 1, T. Nejat Veziroglu editor, pp. 275–283.
16. ibid, pp. 319–334..
17. Kraushaar and Ristinen, *Energy and Problems*, pp. 215–217.
18. Steinhart, Carol and John 1974, *Energy: Sources, Use, and Role in Human Affairs*, Duxbury Press, North Scituate, MA, p. 123.
19. Kraushaar and Ristinen, *Energy and Problems*, pp. 223, 224.
20. ibid, pp. 227–229.
21. ibid, p. 229.

CHAPTER 4

CONVERSION AND END-USE

Some energy resources are suitable for direct use while others demand preconsumption processing to another form and/or conversion to a secondary energy carrier. Often, the character of consumer demand determines the conversion and processing needs. Fluid combustion fuels are preferred for vehicular applications and for most heating needs, where flow rates and burn characteristics can be tightly controlled. Small-scale combustion uses generally require clean-burning fuels, while some large-scale applications can rather efficiently remove pollutants from *dirtier* fuels after combustion. Electricity is absolutely required for electronics applications and is preferred for lighting.

COMBUSTION FUELS

COAL

Nearly 90% of U.S. coal consumed domestically is used to generate electricity, while approximately 10% is used in heavy industry for refining metals, making cement, etc.[1] Since the middle of the 20th century, the direct use of coal for heating homes or for cooking has become uncommon in most industrialized countries because of its smoke and need for handling at the point of consumption. Large-demand consumers can invest in expensive equipment to mitigate airborne emissions. The equipment allows them to burn less desirable fuels, like high sulfur coal, with modest emissions. Individual consumers using less sophisticated pollution control equipment produce more pollution. This is one way in which the environment benefits from large-scale, centralized power generation.

Coal naturally replaced firewood in basic delivery of heat for space heating, cooking (to some extent), and basic industrial tasks such as smithing. The industrial revolution, however, brought coal into a prominent position among energy sources. Coal supplied energy directly to steel-making, to home heating, and to other tasks that required heat. Newcomen's introduction of the steam engine in 1705, and subsequent improvements to the steam engine, then provided a motive force to the industrial revolution. Steam engines converted chemical energy in firewood or coal to heat and thence to mechanical kinetic energy. Mechanical energy produced from external energy sources was not new; wind and water mills had been in use for centuries, but their use was confined to the location of the mill and to the natural energy flux to be tapped. The steam engine allowed industry to apply quantities of energy to a task, limited only by the size of the engine and the quantity of fuel delivered to any chosen task site.

In early steam engines, as steam expanded into a chamber, it drove a piston to the far end of the cylindrical chamber. The moving piston turned a crank to provide mechanical work. In terms of mechanics, these systems were quite like the familiar internal combustion engine. However, the fuel was commonly burned in a firebox beneath a water-filled boiler, and the heated water rather than the combustion gasses of the fuel provided the working fluid in the piston cylinder. The water was heated to its vapor phase before entering a valve at one end of the piston cylinder. The steam's pressure then forced the piston down the length of the cylinder, converting the thermal energy to mechanical work. The piston was typically attached to a crank arm which translated the linear motion into rotational motion. Once the steam drove the piston to the end of its travel, the steam inlet valve closed and an outlet valve opened, so that the returning piston drove the depressured steam out. In some steam engines, the return of the piston was accomplished by a counterbalance weight that pulled the crank arm the rest of the way around to its starting position, and the piston with it. Once the piston returned, the outlet valve closed, and the system was ready for another burst of steam to enter the cylinder.

In modern times, thermal power plants provide a large share of coal-derived energy to industrial societies. Thermal power plants use coal as the primary fuel to generate steam to drive turbines which generate electricity. Turbines are capable of operating as a *closed system.* Steam is recollected at the outlet of the turbine, condensed back to the liquid phase, and returned to the boiler. The mechanics of the turbine are favorable, compared to the piston system, because energy is not expended to drive the piston back to its starting point and expel the spent steam, or in turning a crank arm to convert linear reciprocating action to rotational motion.

The turbine is designed like a fan, with curved blades set on a shaft. The turbine blades are enclosed in a housing which forces fluids passing through the housing across the turbine blades. As steam (or any working fluid) passes through the turbine, it must push the blades around, imparting rotational movement to the shaft. On the other end of the shaft is a permanent magnet, or a conductor. Moving a magnetic field past a conductor or moving a conductor through a magnetic field induces an electrical current. Since the turbine is constantly rotating, it is continually inducing an electric current. The early turbine electric generators were referred to as dynamos and offered a tremendous efficiency improvement in converting chemical energy to useful work. The early system steam generator developed by James Watt operated at no more than 5% efficiency. Modern steam turbines can achieve 40% efficiency.

In modern, as well as earlier steam-powered electric generators, coal is introduced directly to a fire box beneath the boiler which is a tank to hold the water and the generated steam. The steam is superheated in the modern boiler to carry as much thermal energy as possible to the turbine. The French physicist, Carnot, determined that the minimum possible thermodynamic inefficiency of a heat engine is the ratio of outlet to inlet temperatures in the engine (e.g., steam turbine). That is, subtracting this ratio from one gives the maximum efficiency that any heat engine can achieve. This emphasizes the importance of imparting as much heat as possible to each pound of steam.

Coal Combustion • The actual process of burning coal has evolved considerably. The combustion technology can have a great deal to do not only with the level of pollution emitted by the system but with the system's operating efficiency as well.

Stoker feed boilers are units that burn coal directly. They employ the oldest form of industrial coal use, utilizing mechanical means to deliver coal to the fire. In modern applications, a conveyer belt commonly carries the fuel to the boiler, while earlier applications employed *stoker men* to shovel in the coal. The boiler design allows the use of any coal rank. It operates at the particular combustion temperature of the rank of coal available but needs specific size chunks of coal.[2] Coal can also be fed to the boiler in pulverized form. The crushed coal provides a more uniform burn and thus slightly improved operating efficiency.

Staged combustion permits oxidation at a lower flame temperature, which can reduce nitrous oxide (NOx) formation. This technology also facilitates injection of alkaline solvents during combustion. (This is significant since sulphur compounds tend to be acidic; thus, they react readily with alkalines, forming compounds that settle in the ash rather than being emitted in a gaseous state.) Unfortunately, the alkaline solvent injection only eliminates up to one-half of the original sulphur content.

Fluidized bed combustion has the capacity to eliminate essentially all of the sulfurous and nitrous oxides (SOx and NOx). Two forms of fluidized bed combustion that are considered viable are atmospheric pressure fluidized bed combustion (AFBC) and pressurized fluidized bed combustion (PFBC). The distinction, as their names imply, is simply that one operates at atmospheric pressure while the other uses elevated pressures. By employing higher pressures, the PFBC technology achieves efficient burn and contaminate removal, working with physically smaller units. In fact, emissions control may be improved over the atmospheric pressure units, and PFBC has potential use in conjunction with combined cycle. The higher pressure technology has not been developed to the point of significant commercial dissemination by the 1990s.[3]

Coke Production • Coal is often burned directly in many smelting, mineral processing plants. In steel mills, though, the heat required in the iron ore to drive impurities out effectively is higher than the flame temperatures of most direct coal fires. The coal must be converted to coke, a solid mostly carbon material that exists as a residue from superheating coal which burns very hot. Some coke is found in nature, where abnormal heat has been applied to the coal in situ. In practice, coke is almost always artificially generated by the destructive distillation of bituminous coal. Destructive distillation means that the compounds are chemically altered, while the volatile components and products are driven off. In coking, bituminous coal is placed in an oxygen-deprived brick oven and heated to the point that hydrocarbon chains are broken; the ammonia gases and tarry products are driven off and a very carbon-rich residue is left behind. Oxidized coke burns very hot.

Gasification • Converting a solid fuel to a fluid is highly desirable for a number of reasons. The first is versatility. Practical applications of solid coal consumption have become primarily large-scale stationary facilities, such as thermal power plants or steel mills. The large demand sectors of vehicular fuels and household consumption are generally not practical to address with a solid fuel. Thus, coal liquefaction and gasification become topics of interest in relation to the coal's larger resource base than that of oil and gas.

Gasification is the oldest form of producing a synthetic fluid from coal. It was the primary source of *manufactured* or *town* gas used for urban lighting and cooking in England, western Europe, and New England in the 18th and 19th centuries. If coal is introduced to very high temperature steam in a low oxygen environment, the hydrocarbon chains generally break apart—releasing some hydrogen gas, carbon monoxide from incomplete oxidation, some methane from thermal cracking of the chains, and about 50% carbon dioxide. The mixed gas product has low energy content, varying from 150 to 500 Btu/cubic foot, based largely on the amount of methane produced. This compares unfavorably with the 1,000 Btu/cubic foot energy content typical of methane-dominated natural gas. If coal reacts with steam in the presence of pure oxygen rather than the air mixture, more methane results. To bring the heating value of the product gas up to a level that can compare with (and thus reasonably replace) natural gas, the initial product must react in the presence of a catalyst to combine the hydrogen and carbon monoxide to form methane.[4] This incurs an efficiency cost for the system, even though it produces a more readily usable gas.

Coal gasification should not be confused with producing coalbed methane. The latter taps naturally occurring methane that formed along with the coal and may be trapped in the joints and natural fractures of a coalbed. As mentioned in the chapter discussing acquisition, methane is a hazardous mining problem because it is explosive in extraordinarily low concentrations of air. Thus, producing methane from a coal seam prior to underground mining can offer the greatest cost-effectiveness in mitigating the risk of explosion. Methane is produced through a wellbore as an unconventional source of natural gas. This is an expedient process if the coal fractures are adequately interconnected or the coal is sufficiently hard for induced fracturing, similar to that conducted in stimulating conventional oil and gas reservoirs.

Liquefaction • Liquefaction produces an even higher value product than gasification, largely due to the high demand for transportation fuels. In South Africa, a relatively large-scale liquefaction plant converts approximately 4,400 tons per day of coal into 4,000 barrels of synthetic fuel, plus the energy to run the process (slightly more than half of the coal input).[5] The implementation of the South African process reflects the absence of known oil and gas reserves. In addition, energy self-sufficiency was once necessary due to the uncertainty of imports in view of international embargoes imposed in response to their officially racist policies, though apartheid has since been abolished. Interestingly, the Fischer-Tropsch process they employ in their SASOL plant was developed in Nazi Germany who faced similar sets of pressures. Under such constraints, synthetic fuel production from coal is feasible. During the period from 1973 to 1981, when "oil crises" in the United States drove oil and gas prices dramatically upward, the United States seriously considered developing coal liquefaction and gasification capacities.[6] Additionally, coal slurries can be burned directly as a liquid fuel. China and Spain have established plants using 250,000 and 500,000 metric tons of coal per year, respectively. These are modest amounts in comparison to the coal used by a large power plant in a year, but one aspect that particularly makes coal slurries interesting is that significant quantities of coal are commonly lost in conventional processing. When coal is crushed, very fine particles are undesirable for conventional handling and are often disposed of with the noncombustible materials removed in processing. Since about 10% of the total coal volume is commonly lost this way, and very fine particles are desirable for slurry applications, this technology could be important in reducing coal waste and providing a liquid fuel.[7]

President James E. Carter's administration established the Synthetic Fuels Corporation under the Energy Security Act of 1980. The corporation was short-lived because oil supply on the market exceeded demand by the end of 1981 and prices dropped again. In an unconstrained market, synthetic fuels could not compete. The irony for synthetic fuels, as for other alternative energy sources, is that although many consumers voice an interest in the development of energy alternatives, consumers also object vociferously to the increasing energy prices necessary to allow new energy sources to enter the market.

Besides versatility, another factor which has favored fluid production from coal is the improved ease of processing the fluids to remove contaminants such as sulfides. Current combustion technologies for coal, however, can remove sulfides from the emissions streams with comparable efficiency, potentially rendering this factor insignificant.

Coal liquefaction has only proceeded on a commercial scale when political factors have precluded meeting a nation's domestic energy needs through imported oil and gas. The Fischer-Tropsch process is the only plant operating on a significant scale as of the early 1990s. This process effectively involves taking gasification a step further to recombine the carbon monoxide and the hydrogen molecules.

OIL AND GAS

Natural Gas Processing • Natural gas is normally burned directly in its natural state, though the removal of water and the addition of an odorizing agent generally are included in preprocessing. Removal of water from natural gas is important in regions which experience freezing (or near-freezing) temperatures. There are holes in ice crystals, which is why ice occupies more volume than liquid water. In the presence of methane, the crystals can form with methane molecules fitting neatly into the holes in the crystals. This crystal structure is called a gas hydrate, and due to the neat fit, the crystals can occupy less space than the liquid phase water and gas. Therefore, the elevated pressure under which gas pipelines operate favor the formation of hydrates, allowing them to form at higher temperatures than water would normally freeze. The formation of hydrates in pipelines can severely restrict flow. Thus, in cool climates, drying the gas may be supplemented by the addition of some antifreeze compound (e.g., glycol).

Gas production, though, does often contain heavier (more oil-like) components. It is common to remove these products as higher value natural gas liquids. In the simplest case, the gas merely flows into an expansion vessel in which the liquids have the opportunity to condense out of the gas. The gas stream can also be cooled, causing more of the potentially liquid components to condense. Even constituents that are not liquid at atmospheric conditions, such as propane, can be extracted in a refrigeration or cryogenic gas plant as LPG (liquefied petroleum gas or liquid propane gas).

The least common form of gas processing is called *sweetening*. *Sour gas* is contaminated with sulfur compounds, normally toxic hydrogen sulfide. It is considered sour if it contains more than 1 ppm of hydrogen sulfide (i.e., one part of hydrogen sulfide per million parts gas). Removing the hydrogen sulfide is generally required by pipeline contracts before the gas can enter the line. Sweetening processes can produce elemental sulfur, which can be sold as a useful by-product to offset the cost of sweetening. Fortunately, a relatively small fraction of gas resources is sour.

Crude Oil Refining • Crude oil is almost always processed to form products with very specific chemical and combustion properties. In fact, the separation of oil into constituents of different boiling points was essential to its first significant market penetration. Samuel Kier should perhaps share equal credit with Drake for spawning the petroleum industry. Kier's simple still permitted the fractional distillation of petroleum, and he produced a seminal report indicating that the kerosene product burned cleanly, with a constant, bright flame. Thus, kerosene was viable for use in lamps and even for cooking and some heating applications. It was this finding that prompted promoters to hire Drake to get a well drilled. They knew the possibility existed that larger quantities of the newly marketable oil could be found underground.

Essentially, the production of kerosene from crude oil involved technology not particularly different from *moonshining*. Since petroleum products are varying mixtures of hydrocarbon compounds, they each tend to have different boiling points, enabling them to be separated by distillation. As heat is applied to a vat of crude oil, the lightest, smallest hydrocarbon chains are most readily driven to the vapor phase. If the vapor is collected and run through coils of tubing to cool, much of the vapor recondenses as liquid, which can be gathered at the outlet. In crude distillation, the very lightest, smallest compounds do not recondense but remain in the vapor phase to escape at the outlet. Circulating cool water around the outside of the condensing coils increases the amount of condensation and thus the total yield.

To obtain a consistent product, the evaporation and condensation temperature ranges must be tightly controlled. If the crude is heated rapidly, the temperature rises above the boiling point of the desired kerosene range, driving larger, more complex hydrocarbon chains into the distillate. This produces a fluid that burns with more smoke. If the condensation is extremely efficient, the product may contain too many volatile components, which could cause explosions. These were both the case in the early days of the petroleum industry, as many inexperienced people tried to go into the oil business by setting up stills with little or no control.

The advent of the automobile was the great impetus to developing much more complex refining technologies, with many chemical engineers seeking to maximize the gasoline fraction recovery in modern refineries. However, it is unwise to lose sight of the origins of the technology. Energy development did not start in the west with million dollar, automated drilling rigs and billion-dollar refineries. When a project to assist Third Worlders in finding and developing their indigenous oil and gas resources was first introduced, one university professor claiming energy expertise, scoffed in writing, "What will you do with the first barrel and a half of oil produced in Namibia, ship it to Houston for refining?" One of many things wrong with this observation is the loss of contact with the industry's roots. Very little technology or investment is required to set up a moonshine-type still to distill kerosene from crude oil. Well over a billion people on our planet still rely on firewood and charcoal, and kerosene would certainly meet most of those consumers' needs in low income countries.

In the affluent industrialized world, the internal combustion engine and Henry Ford revolutionized the refining industry by changing dramatically the type of fuel most demanded by consumers. Automobiles created vast demand for a product which would burn well in spark plug engines—a fuel that was not kerosene.

The Refinery • Crude oil entering a refinery first undergoes fractional distillation to separate products by boiling point, which generally equates to separation by density and the length of the carbon molecule chain. With increasing temperatures, increasingly long carbon molecules boil out of the liquid. At less than 90°F, the naturally gaseous components (butane, propane, ethane, and methane) break out of the liquid and are sent to the gas processing plant in the refinery. Straight run gasoline evaporates in the 90° to 220° range. Naphtha distills between 220° and 315°, while kerosene comes out in the 315° to 450° temperature range. A product called light gas oil vaporizes between 450° and 650°. As the temperature reaches 650° to 800°, heavy gas oil products evaporate, leaving the straight run residue which remains liquid at 800°F. Products that evaporate in each temperature range tend to have several properties in common and have common uses.[8]

The volumes of crude to be handled in the refinery dictate against *batch* distillation processes, in which a large volume of fluid fills the still, then heat is applied to drive the volatile portions of the fluid to vapor. Refineries require continuous operation. Therefore, the crude oil flows into the distillation vessel where there are many layers of perforated trays stacked above one another and a heat source at the bottom. The heat causes the oil to begin boiling. Naturally, the lightest compounds tend to boil first. It's not that simple though. The heat cannot be applied evenly, so some heavy molecules boil off at hot points, where temperatures are higher than the average fluid temperature at any level in the distillation vessel. Furthermore, rapidly boiling light molecules can carry surrounding heavier molecules with them out of the liquid phase. In order to achieve very uniform products at each distillation level, refluxing, or cycling molecules through condensation and evaporation several times, is required.

At any given level in the vessel, the liquid fraction tends to fall onto the trays while the vapors rise through the perforations in the trays. Each perforation has a cap, forcing the vapor to bubble outward through the surrounding liquid. Some of the vapor's heat is transferred to the liquid as they pass through each other. In the process, the heavier components condense back out of the vapor phase.

As liquids condense, the liquid level builds up in the tray until it is deep enough to reach the top of a drainpipe (called a *downcomer*) through which some of the liquid overflows to fall to the next lower tray. The fire heating the vessel is at the bottom, so each higher tray has a lower temperature. The fluid still in vapor phase at the top has a relatively low boiling point, while the fluid that remains liquid at the bottom, nearest the heat source, has a very high boiling point. Fluid coming off at each level in between has intermediate boiling points, proportional to the closeness to the bottom of the vessel.

The gaseous products drawn off the top of the first distillation column are hydrogen saturated. (The carbon chains contain no double bonds.) They go to the saturated (*sats*) gas plant. There, they are compressed to approximately 200 psi. At this elevated pressure, the heavier components that were carried along with the lighter ones liquefy; then they can be separated.

Liquid products are much more valuable than gasses, so the gasses go through several stages which are meant to *strip off* liquid or liquefiable products. In an apparatus called the *rectified absorber*, mid-size alkanes (naphthas) are pumped in with the saturated gasses. As the gasses bubble up through the liquid in the plant, the heavier gas components, propane and butane, tend to stay with the liquids rather than the gas. Thus, the gas leaving the rectified absorber is stripped down to its lightest constituents, methane and ethane, which could be liquefied only at very low temperatures or high pressures.

The liquid naphtha entering the absorber is called *lean oil*; once it absorbs propane and butane, it is called *fat oil*. The fat oil goes to another separation tower to remove the propane and butane which can go directly to LPG (liquid propane or liquefied petroleum gas) bottling. However, the iso-butane is particularly useful in other processes, so it is likely to be separated from the normal butane and propane in the tallest column in the refinery. (The butane separation tower is the tallest because iso-butane's physical properties are so close to those of normal butane, separation is difficult.) The lighter gases that come off the top of the rectified absorber are likely to carry some heavier molecules with them, so the light gas stream is put through another process, *sponge absorption,* in which a heavy lean oil is pumped through to pick up the valuable heavier molecules.[9]

The heaviest compounds, with very high boiling points, are very long carbon chains. The very long molecules form a thick, viscous fluid or semisolid, such as greases and tars. The need for heavy, thick tars and greases is much less than for mid-range compounds, so it is desirable to break down long molecules into smaller ones. The long molecular chains can be broken or *cracked* by adding sufficient heat and even by mechanical forces. Thermal or mechanical cracking is a relatively random process because the chains can break anywhere. Since uniform products are sought, a more sophisticated process is employed to break the molecules down. The process is called *catalytic cracking*. A catalyst is a compound which enters into a chemical reaction, but comes out of the reaction unchanged. The catalysts interact with the splitting molecules which causes them to break more uniformly. Oil molecules adsorb on the surfaces of the cracking catalysts where it causes some of the oil molecule bonds to dissociate in order to form weak bonds with the catalyst. The weak catalyst-oil bonds break easily to form the new desired bonds and yield shorter molecules. The molecular geometry of the catalyst's surface is important; the reactive atoms of the catalyst should be spaced appropriately for breaking the carbon chain bonds at the desired lengths.[9]

Cracking processes became important to meet changing demand profiles early in the 20th century. W.M. Burton was issued a significant patent for a thermal cracking process early in 1913; in the same year, the Haber-Bosch ammonia catalytic cracking process commenced. Two years later, A.M. McAfee of Gulf Refining Corporation discovered catalytic cracking, with aluminum chloride serving as the catalyst. Catalytic cracking potential of clay residue from motor oil processing was also identified. It was, however, not until 1936 that the first commercial-scale catalytic cracking unit was installed. The Second World War, with its demand for large quantities of high octane fuel for airplanes, provided the impetus for proliferation of cracking to increase the gasoline yields.[10]

Particle and bead catalyst beds were found to be superior. Particles or beads are both highly porous, small configurations of the catalyst. The petroleum products circulate around the catalyst, lifting the small beads or powder-sized particles into a suspension. The pores in the material provide a high surface area for the catalyst to contact the fluids, and the suspension created permits movement of the fluids past and around the catalysts.[11]

In the cracking processes, some small, gaseous molecules are unavoidably broken off, which often contain some double bonds (i.e., they are undersaturated). These light ends go to the cracked gas plant. The lightest unsaturated gas generally is sent to the refinery fuel stream. Heavier cracked gasses undergo alkylation. The alkylation process combines the propylene or butylene with isobutane in the presence of sulfuric or hydrofluoric acid to form (branched chain) iso-alkanes. The gasses are chilled first, so that they can react in liquid form with the acid catalyst. The reaction is relatively slow, requiring a 15- to 20-minute residence time in the reactor vessel. Then, the acid must be separated, and the product must undergo a caustic wash to remove trace acidity.[12]

Another form of catalytic cracking is hydrocracking. The process is different primarily in that it is conducted in the presence of an abundance of hydrogen, which rapidly saturates the broken bonds to minimize the formation of smaller and straighter molecules. This can be a swing process at a refinery, which means that it can be used to generate a higher fraction of gasoline when needed.[13]

Long, straight alkanes are less desirable than cyclic, branched, or even aromatic compounds. Thus, the straight alkanes, called naphtha, are commonly sent to a catalytic reformer which works much like a catalytic cracker in reverse. Alkanes are combined into branched chain and cyclic alkanes. The cyclic alkanes form aromatic compounds. Unfortunately, some cyclic compounds crack into straight chains and small side chains crack off, decreasing the efficiency of the process.[14]

Hydrotreating is a process employed to clean up the liquid products. Hydrogen is introduced with a catalyst to react with sulfur compounds to form the highly toxic and corrosive hydrogen sulfide. Since hydrogen sulfide is a gas, it can easily be removed from the liquid. Nitrogen compounds tend to react with the hydrogen to form ammonia, which can also be drawn off as a gas. The ammonia is useful as a fertilizer, while the nitrogen compounds would increase the nitrous oxide toxic emissions if left in the fuel during combustion. Metals that might be carried in the crude stream tend to be deposited on the catalyst. Side reactions include a tendency for some molecules to become more saturated in the presence of the hydrogen and for some short chain cracking to occur, liberating more methane, ethane, propane and butane—an undesirable reaction.[15]

Kerosene goes to a hydrotreating facility to increase its hydrogenation, or saturation. In this frame of reference, a saturated organic compound is one in which all four bonds in each carbon atom are bonded either to a hydrogen atom or form a single bond to another carbon atom. Many of the molecules that boil in the kerosene range tend to have double carbon bonds that can be broken and one or both bonds replaced by bonds with hydrogen.

Light gas oil goes directly to heavy distillate fuel blending. This yields products such as heating oil and diesel fuel (which are often the same product). The diesel fuels can be blended from the range of characteristics of the alkanes (paraffins) that tend to dominate the straight run light gas oils (those that come straight from the distillation vessel). They have high cetane numbers. Light gas oils that result from cracking tend to be double-bonded alkanes, with lower cetane numbers. The same products are desirable for heating fuel because they have a high energy density but are light enough to ignite easily (unlike the heavier residues.)[16] Since diesel fuels can be produced in quantity from products derived directly from petroleum distillation, there is less environmental impact and less energy lost in the refining phase for diesel and kerosene products than for gasoline.

Heavy gas oil goes to catalytic cracking. The long carbon chains which make it *heavy* are broken into smaller molecules. The process is designed to break them down into *mid-range* products (molecules with about 6 to 10 carbon atoms). The mid-range product generally goes into blending for gasoline. The lighter ends that incidentally break off are sent to the gas plant.

Fuel Requirements • Again, most of the refinery processes in the United States are designed to maximize the gasoline output. Therefore, the requirements of the gasoline fuel are very important to the refining processes. In a traditional gasoline engine, the carburetor mixes gasoline with air to introduce an explosive mixture to the engine in a vapor/aerosol state. In order for the spark plug to ignite the mixture, at least 10% must be in the vapor phase. If it is, the rapid ignition, or explosion, of that fraction of the fuel will promptly ignite the liquid portion successfully, driving the piston to turn the engine. So, it must be sufficiently volatile to meet the minimum vapor requirements during a cold start. Yet, if it vaporizes too readily, the gasoline vapors tend to exclude air from the carburetor when the heat of a running engine vaporizes more of the gasoline. Without air, the fuel cannot ignite. So, the gasoline mixture leaving the refinery for the market must include adequate quantities of volatile and less volatile compounds.

At the same time, the gasoline must meet the familiar octane requirements which are intended to prevent engines from knocking. Engine knock is a relatively serious problem. It is caused by the explosive air/fuel mixture preigniting. The heat caused by compressing the gasses in the cylinder can cause the mix to explode before the piston reaches the top and the spark plug fires. In this case, if the piston has not reached the top of its stroke, the explosion is actually working against the engine—wasting energy and increasing engine wear. Different compounds ignite at different temperatures produced by different compression ratios. Iso-octane is taken as the ideal; it is relatively resistant to preignition and is assigned an octane rating of 100. Normal heptane (with only one less carbon atom) is taken as the opposite extreme, and given an octane rating of zero. The compression ratio at which any gasoline ignites without needing a spark can be measured and ranked on a scale between the two standards. Mixing can be, and is, designed to improve the octane ratio, as needed.[17] Cars with high compression ratios and high power output require high octane fuel. In lower compression, lower horsepower engines, octane number is irrelevant. The consumers can also consider the compression ratio of his or her vehicle in determining what kind of octane rating is required.

Tetra-ethyl lead, now largely eliminated from most fuels due to the toxicity of lead compounds, was once a favored additive. It increased a mixture's resistance to preignition, without changing its volatility or other properties.

Diesel engines have essentially opposite requirements to gasoline engines. Diesels have no spark plug but rather depend on ignition from compression. They operate at higher compression ratios and inject the air before the fuel which enters at the top of the piston stroke where the heated, compressed air ignites the fuel spontaneously. So, the timing of the injection is crucial. Cetane numbers are used for diesel rating rather than octane numbers. Cetane, an alkane molecule with 16 carbon atoms and 34 hydrogen atoms, is taken as the standard. Regular diesel fuels generally have cetane ratings between 45 and 50, with premiums rating between 50 and 55. The premium fuels burn with a little less smoke and are easier to start cold. The mix generally consists of alkanes, cracked and straight run light gas oils, and kerosene. The breadth of compounds that can be mixed allows diesel to be made from most any range of surplus refinery products.[18] Diesel engines operate more efficiently than gasoline engines. A consumer can expect to get something near 20% better fuel economy in a diesel-powered car than in a similar car with a gasoline engine. Diesels have gained considerable popularity in industrialized nations other than the United States. In spite of their efficiency, diesels have visible exhaust emissions, creating the perception that diesel fuel is more polluting. The actual pollution comparison is complex and not definitive.

The design of gasoline for use in the internal combustion engines is very sensitive. The gasoline engine's carburetor mixes the gasoline with air. The air/fuel mixture is injected into the cylinder, the piston compresses it, the spark plug ignites the explosive mixture, and the resulting explosion drives the piston back to the other end of the cylinder, producing mechanical work. The production of mechanical energy by the piston is similar to the process described for early steam engines. The tricky part of a gasoline engine is that the fuel must be volatile enough to enter the cylinder in vapor phase but not explode under the increasing temperature associated with being compressed before the piston reaches the limit of its compression stroke and the spark plug ignites it. This preignition is the cause of knocking.[19]

By no means is all oil used for the production of energy. Approximately 14% of oil in America goes to petrochemical feedstocks to make synthetic products such as plastics, nylon, butyl rubber, etc., asphalts, and lubricants. Thus, of every 42-gallon barrel of oil, only about 35 gallons actually go to energy production. Some experts have suggested that the synthetic products to be made from petroleum are so valuable that it is a shame that we burn any of it. The values placed on synthetics as compared to fuels in the marketplace have not yet substantiated this observation. In the long term, though, when oil resources are effectively depleted, synthetics are likely to remain important, while other sources take over the energy market.

KEROGEN

In its original state, kerogen is made up of more rock than of combustible compounds. It would make a very inefficient fuel if not substantially processed prior to use as a fuel. The processing of kerogen-rich oil shales consists of crushing the rock and then subjecting it to destructive distillation. Disaggregating the material is essential in order for fluids that are generated in the process to be extracted effectively from the rock matrix.

Destructive distillation is a process in which a solid is heated in a closed vessel such that some chemical changes are produced as well as changing the solid to liquid and liquid to gas. The process is similar to thermal cracking as described for petroleum refining. The large complex molecules are broken into smaller carbon chains that form a liquid of similar consistency to a common crude oil. The synthetic crude contains aromatic organic molecules in which the carbon chains make rings, with half of the carbon-carbon bonds double bonds. The synthetic crude can be pumped in pipelines along with petroleum and can enter the refinery, but, if a large share of crude entering a refinery were synthetic crude, the dominance of aromatic compounds would disrupt the balances for which the refinery was designed. Since oil from kerogen has not become significant, neither has a problem at refineries; however, since oil shales seem to have a much larger resource base than oil and gas, it is possible that large-scale kerogen exploitation could create severe stresses on refinery processes. Processes such as catalytic cracking and reforming can ideally take any set of organic compounds and convert them into any desired set of organic compounds. Therefore, the refinery output from synthetic can be essentially identical to the output from petroleum, if kerogen production were ever to become substantial.

BIOMASS

The classic end-use of biomass is direct combustion, which will probably dominate this energy sector until the end of the century. A wood fire was probably humanity's first external energy source. In addition to maintaining market dominance in nonindustrialized societies which depend on it, firewood maintains a romantic mystique in industrialized societies. More subtle than the image of romance in the warm, flickering glow of a fireplace is the notion of wood fires as natural—and thus good and even environmentally benign. This romanticism is significant, for the pervasive notions and emotions generated by the use of wood must be addressed, especially in terms of romanticized, subconscious imagery versus informed and conscious reality.

Granted, as an American, it's hard to imagine a crackling fire as anything else but an ideal setting for courtship. Nevertheless, the author's casual observations have found this feeling much less prevalent among African scholars who have resided for years in industrialized northern countries. It may be that the harsh, smoky realities of firewood dependence leave no room for a nostalgic attachment to the wood fire.

Uninformed observers sometimes argue that dependence on firewood is culturally dictated. They say African women walk 20 kilometers to receding forests to gather a month's supply of firewood because of desire. The fact is that when energy alternatives that clearly meet consumer needs are offered, the transition from firewood proceeds quickly, with less than anticipated persuasion required.[20] Gathering firewood is arduous and time-consuming. In many places, adults work from dawn until dusk in the fields, and the absence of women from the fields is not beneficial to anyone.

Wood is estimated to be the chief provider of domestic energy for half of the world's populace.[21] These people are concentrated in the lower income countries where 70% of the people rely on firewood and charcoal. Climate and firewood scarcity determine the consumption ranges which vary from a minimum of 350 kg per person per year to a maximum of 2,900 kg per person per year. The average Third Worlder is estimated to consume 700 kg per year.[22] With some 2.5 billion people using firewood at the average firewood consumption rate, approximately 1.75 trillion kg or about 2 billion tons of dry wood are burned to provide basic energy needs every year. The number of people dependent on firewood can be expected to increase with the large population growth rates in the nonindustrialized countries, while the forest reserves which supply the wood are being depleted at an ever increasing rate.

A majority of energy demand in lower income countries is for cooking—a very inelastic demand. Frequently, biomass fuel is consumed directly in a simple open fire. When sizable logs of a few inches in diameter and several feet in length are obtained, one end is perched on some rocks and ignited with kindling. After the meal is cooked, the burning end is extinguished so that the log can be slid forward and reignited for the next meal. Since the fires are very smoky, in many areas meals are prepared in a cookhouse separate from the living structure.

Smaller pieces of biomass are piled and burned directly under the cooking vessel; these are similar to kindling in an American or European fireplace. The smaller biomass fire is not preferred because its low density is a disadvantage. A great volume of grasses and twigs must be gathered and carried to the point of consumption. As preferred biomass (generally forest) reserves deplete, less preferred fuels find their way into the fires. An endpoint selection is dung. In this case, animal droppings are gathered and plastered against a wall or rock to dry in the sun. Once dry, they burn fairly well.

An interesting example of the disconnection of many industrialized world citizens with the realities of subsistence living can be seen in the opinion professed by one U.S. government official who believed that burning dung was a "wonderful" option that the consumers enjoyed. Perhaps his perplexed stare that greeted the question of whether he would like the job of gathering the dung is not surprising. Much more significantly, though, he was not considering the environmental consequences of burning fertilizer to cook meals, which will be discussed in the next chapter.

People in biomass dependent regions who cannot gather fuelwood but have money can buy it. Enterprising youths pedal bicycles out to the forests return with surprisingly large loads strapped on the back of their human-powered vehicles. Surveys of the actual quantities of biomass used are very sparse; less available are breakdowns of how much wood is gathered by the consumer, purchased in its original condition, or purchased as charcoal. All of these variations still commonly lend themselves to end-use over an open fire. However, small pieces of biomass and charcoal do have potential for use in *improved cookstoves*.

Preventing a stove's heat from radiating in all directions is a significant engineering improvement. This applies to all types of stoves. The efficiency of transferring energy from the fire to the task of cooking is likely to be 8% or less for an open fire. A simple metal or mud stove that largely surrounds the flame can easily improve the efficiency to 30% or even more. However, research suggests that the laboratory-determined factors that caused the promulgation of *improved cookstove* aid efforts may be misleading. For one thing, the actual efficiencies of cooking over a properly constructed traditional fire with three stones to hold the pot can have higher efficiencies than estimated, and the improved cookstoves may not be used as anticipated. In review of some of the stove projects, the relatively more affluent community members got the stoves and often used them for additional tasks, not for decreasing their firewood consumption. Thus, one can question whether these projects have actually achieved their objectives of reducing fuelwood demands at all.[23] Furthermore, some of these well-intentioned interventions fail to consider the context of consumption. Improved stove design must accommodate the cooking vessels used, which are round-bottomed in many regions. The stoves also may not be usable where whole logs are burned as described earlier. Cutting the logs into small pieces to fit inside a solid-sided stove might be prohibitive.

Needless to say, the direct combustion of solid biomass is prohibitive for most industrialized world energy demands, including vehicular transport and even cooking.

Conversion to Secondary Fuels • The most basic conversion of woody biomass is the production of charcoal. Traditional charcoal conversion, as described previously, is a very inefficient process but yields a much more efficient and convenient product for the end-user. The energy cost in transporting the more massive, unaltered woody product is not a large factor, but the bulk is significant to the merchant handling it. Bulk is also significant to the women purchasing fuel in the markets because a coffee tin filled with charcoal can meet the same cooking needs as a large, unwieldy log or bundle of sticks. The charcoal fire is easier to control, produces less smoke, and can be used readily in more modern, higher efficiency stoves.

In the industrialized world, of course, charcoal consumption is generally discretionary—used most often for barbecuing. Therefore, it is not a significant energy source. The conversion processes employed in that context, though, are much more efficient. Kilns that offer efficiency in producing charcoal for American weenie roasts do not seem promising in the fuelwood-dependent regions because they add to capital costs and are not readily portable. Charcoal production in most firewood-dependent regions must be able to follow the receding forests.[24]

The conversion of biomass to a secondary form has some considerable potential. In the industrialized world, conversion to a liquid fuel receives rather substantial attention due in large part to the preponderance of transport sector demand. In the United States, since our agricultural production is next to none, it is occasionally proposed to use our surpluses of corn to produce fuel. Fermentation of biomass in the presence of yeast yields alcohol, as has been known for millennia. Corn can be fermented and distilled to produce a rather high-proof alcohol (as whiskey makers have long known). One difficulty with this proposal is that the impressive American achievements in agricultural productivity are based on a massive energy subsidy. That is to say that American farming has become very energy intensive.[25]

Another version of biomass conversion is anaerobic digestion, which does offer a viable opportunity to use animal (including human) excrement as an energy source. This option is superior to conventional dung fires because the gas produced (methane) burns cleanly, efficiently, and odorlessly, while the residue in the digestor remains an excellent fertilizer. Collection is a primary limiting factor in the deployment of this technology. Generally, it would be viewed as practical only in the case of a significant number of animals that do not have free range. There are a number of cases where this would be true: feedlots for beef cattle, stables, poultry farms (even free-range chickens or turkeys occupy a relatively small area where their excrement can be readily gathered), hog farms, and perhaps human communities.

Biogas digestion involves gathering the excrement in some container which is sealed to keep air (oxygen) out. Bacteria are introduced which, in the absence of oxygen, digest the organic matter, giving off methane and carbon dioxide as their primary waste products. Nature takes care of introducing the bacteria, which are ubiquitous. The digestor must simply be designed to provide an environment to optimize their growth and activity. Bacteria require organic material, water, and an oxygen deprived (anaerobic) environment. The bacteria thrive and digest the organic material most rapidly in a moderately high temperature. The gas produced can be accumulated in a storage vessel or delivered directly to a combustion appliance. If demand is relatively constant, storage does not need to be an issue, but for most small-scale applications such as cooking, the demand is sporadic and storage can become a major cost factor in the system, because methane is of such low density unless significantly compressed or cooled. The heating value of the gas (per unit volume) is rather low because of the carbon dioxide. This means that the gas cannot be used in a internal combustion engine without significant processing. However, if the source is large enough to justify the installation of an electric generating turbine, it is possible to design such a system to work with low Btu fuel in the boiler.

The process is relatively efficient since it converts about 50%

to 70% of the original chemical energy to methane chemical energy. It has been employed for decades in municipal sewage plants where human excrement is gathered for sanitation purposes. The methane that naturally evolves from settling pits is gathered and used to supply the energy requirements of the sanitation facility.[26] It is almost certain that additional potential exists for the utilization of animal wastes that are concentrated under normal conditions. It may be possible that considerable additional potential also exists to utilize human waste and other organic debris.

Human refuse offers another energy source that is predominantly biotic in its near term origin. In fact, in the United States, the municipal waste stream consists of approximately 56% paper, 9.2% food waste, 7.6% plant waste, and 2.5% wood debris. Of the remaining 19.5%, plastic (generally derived from petroleum) makes up 3.5%, meaning that a total of 84% of the waste stream that is conventionally destined for the dump is combustible.[27] Recycling efforts may change this somewhat; however, noncombustible metals and glass tend to be higher value products, and thus may well be extracted from the waste flow in larger proportion than combustible products.

One option for using the chemical energy content of waste is the separation and direct combustion of refuse-derived fuels in electric power station generators. The problem with direct combustion is that the composition of a garbage stream changes from day to day, making the boiler design problematic to accommodate a constantly changing fuel mix.

To obviate this problem, refuse-derived fuels can be developed. Ferrous metals are easily separated from the stream by magnetic means. Glass and non-ferrous metals are considerably more dense than the organic waste; therefore, passing the material over vibrating grates permits modest-sized dense material to fall out of the stream. Air blasts applied to debris falling through a separation vessel blow the lighter debris across, while large inorganic objects tend to fall to the bottom. Through applications of a series of such separation techniques, a relatively pure combustible waste stream can be achieved.

The organic material can proceed to a shredding facility. Counter-rotating hammer flails separate the larger materials, leaving a more uniform-sized aggregate. Mixing can be intentionally introduced before the material is pressed into relatively uniform briquettes or pellets. These efforts can facilitate the generation of refuse-derived fuels that can reasonably be used in a commercial heating process, such as the fuel for a conventional-type thermal power plant.

If the waste stream is of relatively homogeneous composition, the separation steps can be foregone. Sawmill or pulp mill waste is a particularly good example. Sawdust can be pressed into very uniform pellets, which burn much more efficiently than raw firewood. This form of processed biomass is beginning to serve a significant function within the relatively insignificant niche of direct biomass fuels in the United States. Some energy researchers in non-industrialized countries, such as Ghana, are considering rescaled means of utilizing their lumber-industry wastes. The total impact on national energy consumption would be much larger in lower income countries. In any case, the added potential for such resources is somewhat limited by the fact that they have long been used onsite to generate energy for the mills.

Yet another option to tap the energy potential of refuse is to take advantage of the natural microbial digestion of organic wastes in landfills. The ubiquity of methane-producing bacteria means that methane is generated as the waste is buried in an oxygen deficient environment. Shallow wells can be drilled into old waste dumps to extract the *un* natural gas. A number of wells in dumps surrounding Washington, D.C. produce fuel for the nation's capital in this manner. While this cannot extract all of the original organic chemical energy content, it can be a very practical approach to tap methane that would otherwise escape to contribute to the greenhouse effect in the atmosphere.

If all of the Unites States' municipal landfill combustible matter were tapped to produce energy, the fuel produced could account for as much as 20% of the nation's electric generating capacity.[28] While this 100% use is unrealistic, it demonstrates the potential, and approaching full utilization is not implausible in the larger markets of urban waste, which would still approach 10% of the nation's electricity generation.

Finally, organic wastes can also be tapped through heat recovery in the incineration of toxic wastes. This topic will be considered in greater detail in the following chapter because its total potential for contribution to an industrialized nation's energy production is negligible. Rather, the questions focus almost entirely on the net environmental effects of incineration verses other options. If incineration is desirable, it would seem ludicrous not to tap the energy released for some productive purposes.

NONCOMBUSTION SOURCES

SOLAR POWER

Thermal Applications • Solar space heating can be accomplished with passive systems, which simply take advantage of the natural interaction of visible light with matter. An object's surface appears dark because it is reflecting very little incoming light. Light which is not simply reflected from the surface of matter is absorbed as photons interact with the atoms which constitute the matter. As photons interact with atoms, they impart some of their energy to accelerating the movement of those atoms. On the submicroscopic level considered here, the heat content of a material equates to the kinetic energy of the atomic particles. As the atomic movement decays, the energy is released to the surroundings primarily as lower level infrared radiation—the heat we perceive. This natural process, the conversion of light to heat, is employed in all passive solar space heating designs. It requires no more intervention than to introduce a massive collector with a dark surface to maximize the conversion to heat. Then the heat must only be stored or moved to the point of use, as described in the preceding chapter.

Of course, heat may not be the final form of energy desired but may be a step towards conversion to electricity. In this case, heat is accumulated from the solar radiation incident on the collector to heat a carrier fluid, commonly water, to its boiling point. The resultant steam then is used to turn a turbine to generate electricity. One of the significant points to observe is the number of conversions required in various processes to reach the final energy form, since each conversion step has efficiency losses. For Solar Thermal Electric Conversion (STEC), visible light is converted to heat which produces kinetic energy, turns a turbine, and then generates electricity, which finally performs the useful work. A process with this many conversion steps limits the maximum net efficiency it could ever realize.

Photovoltaic Applications • The interaction of photons with electron clouds causes some of the electrons to become excited, move more rapidly, and potentially jump out of their stable orbital. In the case of a semiconductor material, the geometry of the electron orbitals is such that the electrons that jump end up in a conductive band. If there is a voltage potential across the semiconductor, the electrons will flow which creates a small current. (This process is described in greater detail in the acquisition chapter.)

Solar Thermal Electric Conversion • The natural conversion of light to heat in matter (i.e., the absorption of light by matter) can be put to use to produce steam for use in generating electricity. By concentrating the sunlight, its intensity is increased which is useful in this process. Mirrors can be used to reflect and focus sunlight into a central point or line. At the focal point, the solar energy striking all of the reflecting surface is concentrated to the point that water can be boiled readily.

Another version of STEC is a *power tower* in which mirrors are placed on the ground, aiming their reflected light to a point at the top of a tower, perhaps more than 120 feet high. A plant in Barstow, California using this technology can produce about 10 megawatts of electricity. This is an impressive demonstration, though only 1% of the output of a large commercial power plant.[29]

WIND POWER

Most modern applications of wind power involve conversion of the kinetic energy to electricity. Air moving across the blades causes them to turn. The turning blades spin a dynamo which generates an electric current. (This is described in more detail in the electricity section of this chapter.) The faster the blades cause the dynamo to spin, the more electricity is generated. Indeed, the amount of electricity generated from wind is an exponential (cubic) function of the velocity of the wind. Every time the wind speed doubles, the power output increases by a factor of eight, up to the wind turbine's limits.

Specifications on marketed wind turbines suggest start-up wind velocities ranging from 8 to 12 mph. The average wind speed in Casper, Wyoming is 12 mph. Psychological studies of the residents of Casper have attributed an abnormally high per capita suicide rate to the abnormally high average wind velocity. Whether this conclusion is valid or not, it underscores the reality that high wind areas are not the most amenable habitation sites. Consider, too, that an average wind velocity of 12 mph does not mean that a turbine would constantly generate electricity there. Even Casper has times of virtually no wind to average against the howling 50 mph winds that are well known. While a turbine's theoretical output curve increases exponentially with wind speed, the actual machine has a finite range of velocities within which it can operate. The power output curves in Figure 4-1 shows that small-scale wind turbines on the market are likely to require winds in excess of 3 to 4 mph to start. They *furl* (lose power) at speeds in excess of 30 to 40 mph. When the machine furls at very high wind speeds, it loses its theoretical opportunity to generate its highest electric output.

Nevertheless, wind-generated electricity does have successful applications. For instance, people in nonindustrialized countries or

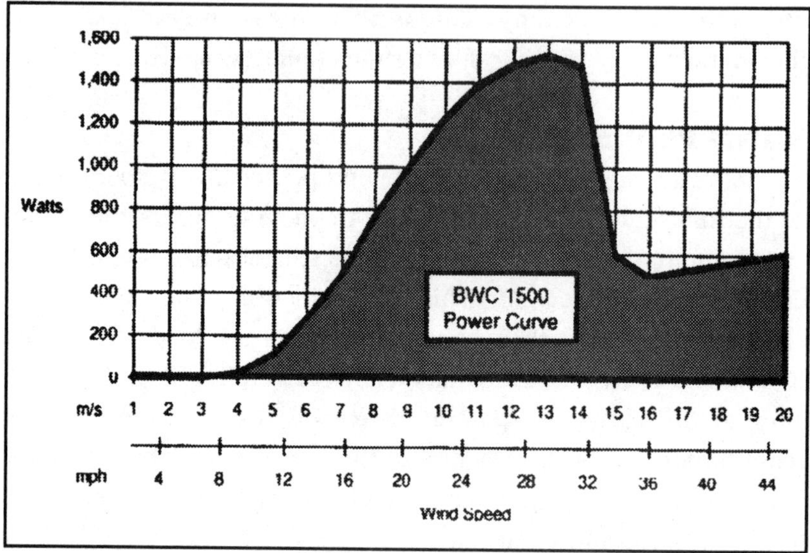

Figure 4–1 The power output curve of a residential scale wind turbine. At wind speeds between 16 and 32 miles per hour, a small-scale turbine can produce more than one kilowatt of energy, but the curve drops off rapidly at higher and lower speeds. A wind machine cannot stand the stresses of wind velocities greatly exceeding those for which they are designed. The rapid decrease in power output at higher velocities is due to a *self-furling* feature of the turbine design represented. Courtesy Bergey Windpower Co., Norman, Oklahoma.

remote settings who can afford the purchase price of wind turbine find it a reliable, serviceable source of power, independent of fuel supply disruptions. In some regions, such as the foothills of southern California, high wind velocities are encountered regularly near populous areas, and some large *wind farms* have been installed that are making dents in the local energy markets.

Wind-generated electricity is a desirable energy carrier but imposes serious inefficiencies, especially when wind speeds are modest. However, the use of mechanical energy produced from wind has been successful for centuries, and technologies that made use of low wind speeds may be worth revisiting because of their equipment's slower, safer rotational velocities.

Slow and Steady • Tasks that require modest, relatively continuous power production have found centuries of wind applications.

Wind has powered textile looms, as well as sawmills and simple manufacturing equipment. Other successful applications include milling grains and pumping water. Thousands of small wind machines can be seen scattered on farms across America. Many of those machines were idled by the advent of electricity which added convenience to the task of lifting water. Since the cost of electricity has been very low in proportion to most American household budgets, convenience keeps those wind pumps idle.

Wind for Transport • Wind was also harnessed for transport for well over a millennium. Several prototype designs for solar-powered vehicles have included some sails to take advantage of tail winds. While this effort has a degree of technological cleverness, it raises rather serious practical questions. How dramatically would traffic flow patterns on industrialized-country highways need to change to accommodate wind-augmented automobiles? Consider the need to raise and lower sails as the powerful drafts of passing and oncoming trucks are encountered. How many drivers would be prepared to keep busy adjusting the sail; it certainly would impair cellular phone conversations. Add to this the considerations that the tail wind must be of greater velocity than the vehicle's and the fact that none of the wind-aided prototype vehicles have performed well in solar vehicle races. The notion of using wind on the highways seems very doubtful.

Of greater practical utility are proposals to resurrect sailing ships. Steam ships did not defeat the sailing vessel industry in general applications until the middle of the 20th century. Time discounting killed the wind-powered shipping industry; delivering the product for sale to recoup expenses in a matter of days rather than weeks is more desirable to most investors. However, sailing ships are less expensive to operate because of massive fuel savings, though these savings are partially offset by additional crew requirements. If combustion fuel taxes or real price due to scarcity rise sufficiently, wind may be able to contribute significantly to the international energy picture.

HYDROPOWER

Most modern hydropower projects aim at generating electricity from the kinetic energy of moving water. They are also typically done on a very large scale to spread the costs of the entire project across a large amount of energy production. Modern river hydropower projects normally involve constructing a dam to retain a great deal of water and to create a large, immediate hydrostatic head. The head refers to the height of a column of water and the pressure exerted at the bottom of the column by the weight of the water above it. The greater the head of water entering a turbine the more energy it can impart to the task of generating electricity. Therefore, dams of more than 500 feet in height are not uncommon. Water is directed to the turbine through large pipes (as much as 30 feet in diameter) called penstocks. The massive stream of water from the penstocks forces the blades to turn as the water flows through the chamber to emerge at the bottom of the dam and reenter the riverbed.

The large dams, in addition to providing a very high head for large energy production, also provide a great deal of storage as water backs up behind the dam, flooding thousands of acres surrounding the original riverbed. The dams often serve an additional flood control function, while the huge reservoirs are considered to offer added recreational benefits. The benefits attributable to large-scale hydropower are offset by the environmental and economic costs of flooding land to form the reservoir and disrupting the flow of the river. These issues, which will be explored in greater detail in the final two chapters, focus some attention on the redevelopment of small-scale hydropower. The principles remain the same as for large-scale plants, but rather than having water heads exceeding 500 feet in height, the small-scale projects typically exploit less than 65-foot water heights.[30] World Bank information suggests that the economies of scale make the electricity produced in gigawatt-size plants cost about one-third as much as electricity generated in very small plants. Building a plant that produces one gigawatt of power does not cost 20 times as much as building a plant that produces only 50 megawatts of power. Furthermore, the installation of power lines and other infrastructure to carry the electricity to the consumers represents a large capital cost that can be spread out over many more kilowatt-hours of electric sales with the large plant than with the small one. Studies, however, do suggest that the loss of economy of scale only occurs for plants of about 50 megawatts and smaller.[31]

GEOTHERMAL POWER

The heat carried from the earth's interior by produced fluids is converted to electricity in steam turbines like those used in combustion fuel power plants. In the ideal geothermal reservoirs, water exists in pores and natural fractures of the subsurface reservoir as super-heated steam. As the steam is allowed to expand through a turbine chamber, it turns the blades of the turbine. The water loses much of its energy in the process of expanding and turning the turbine blades; it tends to condense after it has done its work.

In the case in which the reservoir temperature is not sufficient to vaporize the water completely, a working fluid can be used that has a lower boiling point than water. Since the difference between inlet and outlet temperatures represents the amount of energy that a vapor turbine can extract, there is an advantage to employing a lower boiling-point fluid in the operation. Additionally, the secondary fluid enables the produced water to be kept in a closed loop from which essentially none of it, and the toxic contaminants that may be dissolved in the water, can escape to the atmosphere. However, the heat transfer cannot be perfectly efficient, and environmental problems have been identified with some of the thermodynamically preferred working fluids such as toxic ammonia and ozone-destroying fluorocarbons.

NUCLEAR ENERGY

DIRECT FISSION REACTIONS

A majority of nuclear energy is produced in reactors that induce fission of naturally occurring uranium 235 (U-235) which has been separated from the other much more abundant isotopes of uranium. Fissile uranium splits when an additional neutron enters the nucleus, with a loss of mass and release of a large quantity of energy (about 195 million electron volts per atom, as compared to about 30 electron volts of energy per molecule of TNT exploding). When more than the *critical mass* of fissile uranium is present, the neutrons released by the splitting nuclei are sufficiently abundant and close together to sustain a chain reaction. That is to say that the uranium needs to be bombarded with neutrons to start the reaction, and then the neutrons released in each fission reaction (about 2) are numerous enough and closely spaced enough that the probability of encountering other nuclei and inducing another fission reaction reaches 100%.[32]

Control rods of inert material and moderator materials regulate the reaction. In the most common reactor designs, water serves as the moderator and the heat transfer fluid. Water circulates between the fuel rods. (There are 46,000 fuel rods in bundles of 54 in the common boiling-water reactor [BWR].) As neutrons pass through the moderator, they slow down, losing some of their kinetic energy after their vigorous ejection from the splitting nuclei. Slower *thermal* neutrons are much more likely to interact with the uranium nuclei than at their high energies of emission.

Control rods inserted into the reactor core present obstacles to the movement of neutrons from one set of fuel rods to another. The control rod is made of material with a high neutron capture cross section—which means that in addition to blocking neutrons, it absorbs them. The chain reaction can be slowed or stopped by reducing the number of neutrons moving in the core.

The water moderator in a BWR also serves as the coolant and as the heat transfer medium. Water in the reactor vessel absorbs a tremendous amount of heat from the nuclear reactions and is quickly driven to super-heated steam. The steam flows out of the reactor vessel to a turbine, generating electricity in the conventional means. In the variation used in the smaller reactors of the U.S. Navy, the pressure of the fluid is high enough that it remains liquid. The hot, high pressure liquid flows through a heat exchanger to drive a separate water source to vapor to drive the turbine. This type is called the pressurized water reactor (PWR).

The reactor vessel of a large 1.2 gigawatt (GW) nuclear power plant contains about 155 tons of fuel, which is enriched uranium. The concentration of fissile U-235 is enriched from its natural 0.7% to about 3%. The bulk of the uranium is the non-fissile isotope 238.[33]

ENHANCED FISSION REACTIONS

The U-238 which makes up most of the fuel rods also absorbs neutrons and then undergoes prompt decay to Pu-239, which is also fissile. Since the stable isotope of uranium is so much more abundant, it is possible for a reactor to produce more fissile material (from the non-fissile isotope in the reactor) than it consumes (of the original fissile uranium). A reactor designed to make use of this process is called a *breeder reactor.* Eventually, even with breeder reactor technology, all of the uranium feedstock would be depleted, but estimates are that this would take some tens of thousands of years.

Breeder reactors operate at extremely high power densities (energy output per unit time and mass). This requires very efficient heat transfer fluids. Liquid sodium is proposed instead of water for a reactor design proposed for Clinch River, Tennessee. Liquid sodium transfers heat very efficiently and remains a liquid at very high temperatures, while not tending to absorb many neutrons. These are all desirable qualities for the energy intensities of breeder reactors.[34]

CONTROLLED FUSION

Nuclear fusion occurs when two light nucleii are brought together with sufficient energy. Within the physical limits of our solar system, these reactions are essentially limited to reactions of hydrogen nucleii. The fusion reaction of the lightest nucleii is about eight times more efficient in converting matter to energy than are the fission reactions of very large nucleii. Nearly 1% of the nuclear mass is converted to energy in fusion reactions.

The energy required to initiate the fusion reaction equates to a temperature approaching 50 million degrees Fahrenheit. It is possible to create such high energies by using a fission reaction as a trigger, by particle accelerators, or in the point of a laser. Bringing the nucleii together with sufficient energy to react is the problem. Anything that imparts sufficient energy to initiate the reaction gets the nuclear particles of the fuel very excited. In their excited state, the particles fly apart for the same reason that heating a gas causes it to expand.

Fission triggers surround the fusion material in hydrogen bombs, in which the reaction need not be controlled. When the small fission bombs are detonated, their explosion forces the material in the center together with tremendous energy. Of course, the fusion conditions last for a fraction of a second, before the reactants are scattered in a massive explosion. But sustaining a controlled reaction is much more difficult. First, the reaction in a power plant must be much smaller than in a city-leveling H-bomb detonation. Second, the smaller-level reactions must go on for years rather than nanoseconds.

Therefore, a nuclear fusion plant must incorporate a design that holds the nuclear material together at sustained temperatures far beyond the melting points of any known materials. Some scientists are working on means of using magnetic fields to confine the super-heated plasma, while others work on a concept more like that of fusion bombs, using lasers to provide the intense energy from several directions at once.

Nuclear fusion reactions, though very prompt, do require some time with sufficient particle density and temperature. The particle density and time requirements are combined in the Lawson criterion, that the time (in seconds) multiplied by the particle density (in number of particles per cubic centimeter) must exceed 10^{14} sec/cc. If the hydrogen particles could be held together at the density of air and the threshold temperature for one hundred thousandth of a second, the criterion would be satisfied. It is possible to generate controlled nuclear fusion reactions in the laboratory, but the techniques employed use more energy than the reactions release thus far.[34]

ELECTRICITY

GENERATING ELECTRICITY

Michael Faraday is credited with the discoveries of electromagnetic principles requisite to the development of dynamos to generate electricity from mechanical energy and to produce mechanical energy from electricity. His discoveries around 1830 coincided closely with discoveries by Joseph Henry, but Faraday was the first to publish his findings, winning his place in the history of energy developments.[35]

A dynamo is a common form of electric generator. It uses the effects of magnetic fields moving relative to an electrical conductor. A conductor is a material with valence electrons that are free to flow from one atom to another. If a conductor is moved through a magnetic field, the force of magnetism pushes the electrons to the side, creating an electric current. The current lasts only as long as does the conductor's movement across the field. In order to maintain a current, it is necessary to keep the conductor moving relative to the magnetic field. Hence, the principle of turbines is to use the mechanical energy of moving fluids to spin a dynamo, which can hold either coils of wire or magnets. In a typical case, the portion of the dynamo spun by the turbine blades (the armature) holds wire coils, which are surrounded by magnets, whose fields are constantly being cut by the conductors. (If the spinning portion holds magnets, the principle is the same in the converse.) Thus, a portion of the mechanical energy is converted to electrical energy.

Most commercial, large-scale systems are designed to produce electricity in the form of alternating current (AC). The armature spins through magnetic fields on either side. The motion of the armature creates electron flow in one direction on the upward side of its spin and in the other direction as the armature passes downward through the other side of the magnetic field. To generate direct current, the geometry of the armature passing through the magnetic field must be arranged to maintain a constant direction.

A transformer can be used to change the voltage of a current. Before entering a long-distance main power grid, the current is normally run through a *step-up transformer.* Long-distance transmission requires high voltages to permit the movement of vast quantities of electricity through wires of modest diameter, as discussed in the previous chapter. Before it reaches the consumer, though, the current's voltage is decreased to safer levels by passing through a *step-down transformer.* Transformers employ no moving parts but rely on electromagnetic induction. The movement of electrons through a conductor creates an electromagnetic field around the conductor. If another conductor is nearby, the *moving* field of alternating current in the original electrified circuit continually induces (creates) a current in the nearby, secondary circuit. The voltage in the secondary circuit is proportional to the voltage in the primary circuit. It is also a function of the number of times wire is wrapped around the coil in the primary circuit and in the coils of the secondary circuit. That is, if the secondary coil has twice as many loops as the primary, its voltage will be twice the primary voltage, or half if it only has half as many loops as the primary. This phenomenon is employed to adjust voltages easily.

USE OF ELECTRICITY

Electrical energy is extraordinarily versatile in its applications because of the range of changes an electric current produces in different conductors. A current passing through a conductor produces temporary changes, moving electrons out of their orbits, which can release some of the energy in varying degrees and ranges of the electromagnetic radiation spectrum. A tungsten filament in the near vacuum of a light bulb emits a large proportion of energy in the visible light spectrum. The electron flow of current ionizes gasses in neon and fluorescent lights, releasing an even larger proportion of the total energy as visible light. Other filament arrangements emit a large percentage of energy in the infrared (heat) range of the spectrum. The recognition that current could pass through selective elements, polarizing them, led to the advent of the early binary computers. The miniaturization of circuitry has led to an incredible boom in sophisticated computing devices with an impressive range of capabilities.

Electricity can turn motors in exactly the opposite process to its production in a dynamo. This process is employed to operate such appliances as fans, compressors for refrigerators and air conditioners, and all sorts of motorized equipment. These applications can be met by combustion fuels, and in some cases more efficiently, but with less versatility in the home or workplace. Building codes in the United States call for electrical outlets frequently spaced (about every six feet along the walls of living space). The electricity is then readily available wherever one wishes to place an appliance. Whereas it would also be possible to build a house with numerous outlets for natural gas (as they were at the end of the 19th century), more materiel is required for piping, and it would be more difficult to retrofit a new piped outlet than to add wiring.

While it is much less efficient to heat a living space through the infrared emissions of resistive filaments than by circulating air warmed directly by the combustion of fossil fuels, many homes have used electricity for this task and still use it to provide heat for cooking, heating water, and drying clothes. The fine control and uniformity possible with electricity seems to be the largest factor in this trend, a trend towards inefficiency, though, that has begun to reverse since the energy price increases of the 1970s.

CONSERVATION

ABSTINENCE

The most direct way to conserve energy is simply to abstain from discretionary use. This can include the well-publicized measures of walking or bicycling on trips of less than a few miles, turning off lights and appliances when they are not in use, turning down the thermostat, and so forth. For the scope of the present chapter, the issue is the end-use savings involved in reducing discretionary energy use.

Some abstinence measures involving sacrifices of benefits offered by consumption, others really do not. For instance, establishing a comfort level in living or working quarters is the end-use purpose of space heating. Turning thermostats down has the obvious drawback of reducing the comfort level; therefore, the consumer does make a sacrifice in this type of conservation. Conversely, the act of turning off unused appliances does not reduce the comfort or entertainment provided by the appliance. Consumers would probably site time-savings as the reason for using the car for short trips, but the time actually saved can be minimal. It is not unusual to enter a store or restaurant during the summer which is air conditioned to the point of being uncomfortably cold. Why do people consume energy that provides no significant benefit?

It would seem apparent that the cost of consuming the resource is not sufficiently high to demand the consumers' attention. The cost of leaving a television running when one leaves the house is so low that it does not even impinge on the consciousness of many people to bother turning off the switch. The publicity presented by numerous environmental advocacy groups is to raise consumer awareness of the *externalities*: the costs to the environment and society of energy consumption which are not seen in the price the consumer pays for energy.

EFFICIENCY IMPROVEMENTS

The category of energy conservation covered by improvements in efficiency refers to measures to realize the same end-use benefits with less energy. These steps generally require more effort than does simply choosing to forego the consumptive activity but may not involve any loss of the end-use benefits.

In some cases, efficiency measures do incur some loss of benefit, though. The most fuel efficient automobiles are smaller, lighter, and have less engine power than the larger, more powerful gas guzzlers. The consumer does forego the benefits of rapid acceleration, of four-wheel drive, of spacious interiors, and of crash survivability in selecting an efficient car. The consumer also sacrifices privacy and schedule flexibility in opting for public transportation, a sacrifice, though, which is somewhat offset by the gain in comfort and relaxation or work time possible when not devoting attention to the road. Weighing the costs and benefits of consumption choices is a topic for the final chapter.

In terms of vehicular transportation, the internal combustion engine is of innately limited efficiency. Modifications to the components of the engine to reduce internal and inertial resistance, combined with fuel injection systems and higher compression ratios, can afford some improvement. A story commonly surfaces during times of high energy prices of a carburetor that will make an ordinary car get 200 miles per gallon. Kraushaar and Ristenen, in their text, address the implausibility of this mythical carburetor. They itemize efficiency constraints unrelated to the efficiency of the chemical combustion of the fuel, which is the factor influenced by the carburetor design. The sum of energy lost as waste heat is approximately 74% of the chemical energy content of the fuel. While the design of the engine (not the carburetor) can reduce heat loss somewhat, large efficiency losses in waste heat are inherent to the process of using the expansion of gasses to drive a piston, which must compress gas for the next explosion as well as imparting its mechanical energy to the drive train.[36] The efficiency of an electric motor can be so much greater than that of an internal combustion engine that, even considering the efficiency losses in a thermal power plant (on the order of 60% loss), the net efficiency of electric cars can be cost-competitive with gasoline-powered automobiles.

The potential benefit from improved aerodynamic design is limited in comparison to the potential for improvements in energy conversion efficiency, but not insignificant. Vehicular consumption is such a large fraction of total energy use that even modest savings can be more important to the total energy flow picture than large efficiency gains in smaller consumption sectors.

Improving the insulation of a home is one conservation measure that generally enhances the benefits of space heating because a drafty house is less comfortable than a house with little external air infiltration. At the extreme, though, a super-insulated house can have so little exchange with outside air that the air in the living space can become stagnant and filled with carbon dioxide (from breathing) and noxious vapors (from vinyls, paints, and insulation), not to mention cooking odors, smoke, etc. These houses actually require vents that expel stale house air and draw fresh air in from the outside. The vent systems generally incorporate a heat exchanger, so that the incoming air can pick up much of the heat of the outgoing air.[37]

Considerable gains have been realized in the category of industrial energy use since the increasing energy prices of the 1970s. Control technology has played a large role in this by optimizing the air to fuel mixtures and maintaining ideal temperatures of combustion. (The maintenance of optimum air/fuel mixes not only improves combustion efficiency but also minimizes the formation of nitrous oxides pollutants). Such improvements, of course, are functions of the advances in computer technology as well as the increasing energy prices they accompanied. Further gains can be expected as older plants are decommissioned and replaced with plants employing the best technologies.

Cogeneration offers one of the larger potentials for industrial energy savings. Heat is a waste product of most industrial processes. Steam turbines invariably maintain high temperatures in the effluent streams, even after employing a heat exchanger to transfer a large portion of the effluent heat to preheat the inlet water stream. If a heat exchanger is perfectly efficient, both the effluent and inlet streams end up at the same temperature, which is halfway between their starting temperatures if masses are equal.

Thus, the effluent is conventionally released to the environment at a considerably elevated temperature.

Therefore, if that remaining waste heat is diverted to some useful task, such as providing space heating for the plant or a near-by community, a significant improvement in overall energy conversion efficiencies can be realized. Another version of making use of the waste heat in industrial processes is to use it to preheat air used in combustion processes. This allows a more efficient production of heat in the combustion chamber.

ENDNOTES

1. Energy Information Administration 1991, *Annual Energy Outlook*, U.S. Department of Energy, Washington, D.C., p. 44.
2. *Clean Use of Coal* 1985, International Energy Agency, Paris, pp. 15, 16.
3. ibid, pp. 75–86.
4. Kraushaar, Jack and Ristinen, Robert 1993, *Energy and Problems of a Technical Society*, Wiley and Sons, NY, p. 63.
5. Steinhart, Carol and John 1974, *Energy: Sources, Use, and Role in Human Affairs*, Duxbury Press, North Scituate, MA, p. 62.
6. Kraushaar and Ristinen, *Energy and Problems*, p. 62.
7. Basta, Nicholas, et al 1994, "Coal Slurries: An Environmental Bonus," *Chemical Engineering*, McGraw-Hill Inc.
8. Leffler, William 1979, *Petroleum Refining for the Non-Technical Person*, PennWell Books, Tulsa, OK, pp. 49–51.
9. Degering, Ed F. 1972, *Organic Chemistry*, Barnes and Noble, NY, pp. 384–385.
10. Avidan, Amos, et. al. 1990, Innovative Improvements Highlight FCC's Past and Future," *Oil and Gas Journal*, January 8, 1990.
11. Leffler, *Petroleum Refining*, pp. 40, 41.
12. ibid, pp. 59–63.
13. ibid, pp. 81–85.
14. ibid, pp. 66–68.
15. ibid, pp. 113, 114.
16. ibid, pp. 104, 105.
17. ibid, pp. 91–94.
18. ibid, pp. 88–104.
19. ibid, pp. 91–97.
20. Baah-Boakye, "Keeping Ghana Green and Energy-Efficient," *EcoJustice Quarterly*, Winter, 1993–1994.
21. World Commission on Environment and Development 1987, *Our Common Future*, Oxford University Press, NY, p. 15.
22. G. Foley, *Wood Fuel and Conventional Fuel Demands in the Developing World*, AMBIO, vol. 14, #5, 1985.

23. National Research Council, *Diffusion of Biomass Energy Technologies in Developing Countries,* National Academy Press, Washington, D.C., pp. 6,7.
24. ibid, p. 39.
25. Steinhart, *Energy*, pp. 79–85.
26. Kraushaar and Ristinen, *Energy and Problems*, pp. 277–279.
27. ibid, p. 272.
28. Edmonds, Jae and Reilly, John M. 1985, *Global Energy Assessing the Future,* Oxford University Press, NY, p. 231.
    Energy Information Administration, op cit, pp. 44–45.
29. Kraushaar and Ristinen, *Energy and Problems*, pp. 170, 171.
30. Cassedy, Edward S. and Grossman, Peter Z. 1990, *Introduction to Energy,* Cambridge University Press, NY, pp. 33–36.
31. Edmonds and Reilly, pp. 220–223.
32. Kraushaar and Ristinen, *Energy and Problems*, p. 99.
33. ibid, pp. 101–105.
34. ibid, pp. 107, 108.
35. ibid, pp. 108–113.
36. Kraushaar and Ristinen, *Energy and Problems*, p. 318.

CHAPTER 5

THE IMPACT OF ENERGY USE

The term *impact* is commonly used in reference to the harmful environmental effects of modern energy production and consumption. However, the use of energy has both positive and negative impacts on the economy and quality of life as well. It has both benefits and costs. The use of external energy sources is necessary for human survival and for economic and technologic development. This is not to say that the oft seen graphs of national GNP versus Energy Consumption per Capita describe an unalterable relationship. It is to say, though, that a growing population requires an increasing utilization of energy to provide for a thriving society. The utilization of energy, as opposed to the mere consumption of energy, refers to the amount of useful work obtained from the energy consumed. Thus, efficiency gains can increase the total energy utilized just as the addition of new energy input.

Conservation can displace a certain amount of demand. However, the option to displace one energy form is tantamount to opting for the sources not curtailed. For instance, if conservation efforts in the United States accomplished a mere 2% reduction in American energy demands, nuclear power could be replaced altogether. The decision to eliminate nuclear power, though, equates to a decision not to reduce coal or imported oil use. Energy consumption in most industrialized countries comprises a broad mix of energy forms, each with its own set of costs and benefits. Careful analysis should be conducted to determine which energy resources to displace.

Unlike the industrialized world where the total energy consumption pie is large and diverse, most nonindustrialized countries have a small pie with only biomass as filling. Therefore, conservation opportunities are much more limited in those countries than in the industrialized ones because they are largely dependent on biomass and have little discretionary energy consumption. It may be possible, though, that as these countries industrialize, they will have the opportunity to conserve *a priori* by adopting energy efficient technologies.

COMBUSTION FUELS

Air pollution, global climate change, and acid rain are environmental costs attributable to the use of combustion fuels. The solid fuels (coal and firewood) tend to release particulate pollutants when burned while the fluids (oil and gas) generally do not emit particulates or leave an ash residue. Whenever combustion occurs with oxygen provided from the air, some atmospheric nitrogen oxidizes as well. The oxides of nitrogen are significant in the development of smog and are minimized if combustion temperatures are kept low. Internal combustion engines in vehicular transport are major contributors to nitrogen oxide emissions.

GREENHOUSE WARMING

Climate change is commonly referred to as greenhouse warming, but the latter is a misleading term. The real phenomenon relates to the absorption of infrared radiation by certain molecules. Carbon dioxide, methane, and some other molecules tend to absorb short wave-length radiation, much as dark objects absorbing radiation in the visible light spectrum. Since energy from visible light that is absorbed on the earth's surface tends to be reradiated in the infrared range, outbound energy is composed largely of infrared radiation while incoming energy from the sun is heavily dominated by energy in the visible light portion of the spectrum.

That means that gasses that trap infrared energy do very little to reduce the energy influx to the earth but prevent the flux from escaping to space; consequently, the gasses tend to warm the earth. However, since neither the radiation flux nor the emissions of greenhouse gasses are uniformly distributed across the earth's surface, a simple, uniform warming cannot be expected. An equally serious complication in predicting the effects of carbon dioxide released in huge quantities from modern-day societies' combustion fuels relates to feedback mechanisms. The earth is not a static, passive recipient of anthropogenic changes. It is a dynamic system that responds to change with changes of its own. A great deal of debate revolves around the actual effects to be expected and the extent to which the earth can and will respond to carbon dioxide emissions by counteracting or balancing changes. A great deal of carbon dioxide is contained in solution in the oceans, and some argue that the oceans provide an incredibly vast carbon sink that will simply take on more carbon as more is released to the atmosphere and prevent significant atmospheric buildup. Others expect the opposite, that as temperatures begin to change local climates will change, threatening the lives of carbon-fixing organisms and changing water temperatures and currents. They say that the feedback mechanism may actually be for the oceans to give up some of their vast carbon dioxide store to the atmosphere.

If the scientific theories of general increases in global temperature are substantiated, one change that can clearly be expected is a partial melting of the polar ice caps. A global increase of only a few degrees Fahrenheit would produce enough melting of polar ice to raise ocean levels and flood many populous coastal areas. Even if an overall warming is not seen, scientists predict local climatic destabilization. Because the carbon dioxide accumulations are not uniformly distributed, some areas tend to warm more than others. In general, this process is perfectly natural—it is the main cause of winds. However, anthropogenic warming can change natural air flows which in turn affects weather patterns. Areas that once received considerable rain may become dry, and arid zones may gain precipitation. Since the local ecologies have evolved to match their climates, such changes can be disruptive to many species, including humans. For example, agricultural investments have been made based on climate patterns. If humid areas become drier or arid regions become moist, the economic impact on farmers may be very large.

Whether climate change will prove to be a large environmental impact or not, both the release of carbon dioxide and the removal of carbon sinks contribute to the theoretical phenomenon. Since the earth continues to experience massive deforestation, one very large sink is seriously jeopardized. Firewood consumption stands alone in contributing both to the emissions of carbon dioxide and the loss of a sink. While some argue that the emissions of carbon dioxide from burning biomass do not count because they are taken back up in growing biomass, this argument is highly suspect and will be addressed in the final chapter. What is clear is that if global climate change is a serious threat, minimizing carbon dioxide emissions is desirable. This offers some motivation to move away from combustion fuels. The possibility of this threat also suggests using fuels as efficiently as possible.

ACID RAIN

Acid rain is primarily caused by the reaction of sulfur dioxide with water vapors in the atmosphere to produce sulfuric acid. Nitrogen oxides can also produce nitric acid, and carbon dioxide can produce carbonic acids. Sulfuric acid is deemed to be by far the largest contributor to acid deposition (the more general term that includes acid rain and the dry deposition of acidic compounds). Acidification has been observed in the lakes and forests of Europe and North America. It can produce massive fish kills in lakes, and it can endanger trees in forests. The mechanisms that produce tree death are not clearly known, though anyone who has tried to maintain an aquarium has likely experienced the sensitivity of fish to the pH of their environment. The change in pH alters the chemistry of the water and the soil. The pH changes can dissolve chemicals such as aluminum compounds that were previously insoluble in both soil and water. The plants and fish, respectively, may not be killed or seriously weakened by the dissolved compounds entering their systems. Often, the fish or plants are actually killed by common diseases or parasites which their natural defenses would have been able to overcome, if not already stressed by coping with the unfamiliar chemicals entering their systems.

Sulfur is the major culprit in acid deposition; it occurs in relatively small amounts in biomass, and it can be rather easily removed from fluids (oil and gas) in a *sweetening* process prior to combustion. Therefore, this problem will be discussed at greater length as an environmental cost of coal use. Nevertheless, the form of energy consumption is important as well as the composition of the fuel. About one-third of acid rain is believed to result from chemical reactions of nitrogen oxide emissions, and half of those come from automobiles.[1]

COAL

Coal has a significant range of impacts associated with every step of its use, many of which are popularly known and discussed.

Acquisition • The scars on the surface of the land produced by strip mining may represent the most visible impacts and generate the most emotional reaction to coal's use. It is very true that the environmental impact of strip mining without reclamation is egregious. However, modern mining in industrialized nations invariably includes extensive reclamation. The Surface Reclamation Act of 1977 placed stringent requirements on mining companies. These companies must return the surface contours to match the original contours closely, and they must replant vegetation. Erosion of the restored surface while new vegetation is taking root can be a serious problem. Consequently, the companies are compelled to move quickly in re-establishing growth, with a preference for planting fast-growing species. Stands of pine trees in the Ohio River Valley are good indicators of rather recently reclaimed strip mines. While selective planting of such fast-growing species can produce some short-term ecologic disruption, experience has shown that within a human lifetime, indigenous species may reestablish themselves.

To facilitate the reclamation effort, the stripping process starts with removal and piling of the topsoil in a pile separate from the overburden. When reclamation begins, the overburden spoils are first placed in the pit and roughly graded to match the original surface contours. Then the topsoil is replaced and graded. Finally, the selected vegetation mix is planted. The results of strip mine reclamation range from total success to hopeless failure, depending on perspective. What is clear is that reclamation is feasible.

The purpose of devoting considerable effort to restoring the contours, or shape of the surface, is to minimize erosion. The initial planting selection is designed to generate rapid early growth for the same purpose, but also to forestall the complaints of the public in viewing denuded land.

Surface subsidence occurs as overburden settles into the voids created in underground mines. The discussion of pressure gradients of pore-filling fluids from Chapter 2 may provide some useful concepts in understanding subsidence. Under normal lithostatic loading conditions, the rock matrix supports the load of overlying rock while the fluid in the pore spaces supports the load of fluids filling the pore space of the overlying rocks. In such a case, the withdrawal of fluids from a rock stratum does not directly remove support for the surface rocks. However, the removal of rock matrix reduces the structural support. Consequently, the ultimate subsidence of the surface overlying a coal mine is to be expected if no particular measures are taken to guard against it. Measures such as timbering to support the mine roof can be very effective in the near term while the mine is operating but are likely to fail over subsequent years or decades, especially when the support is provided by wood timbers that can decay in the moist subterranean environment. Putting debris from the ongoing operation back into worked-out sections of the mine can minimize the problem.

While a degree of ultimate subsidence is likely, the problem has some natural mitigation. As the immediately overlying rock formation collapses into the mine, it fills the mine with loose rubble. If the overlying beds were well-compacted, the rock grains would have been packed efficiently, with minimal porosity. Thus, the beds would have occupied less total volume than they could if the porosity were increased. The process of collapsing into the mine to form rubble does precisely that. It converts an efficiently packed bed of rock grains into an inefficiently packed, larger volume. So, a point can be reached at which a solid column of rubble occupies the void created, without allowing the surface rock formations to collapse. The overburden load will probably cause the rubble to settle constantly until a comparably efficient packing was achieved. This kind of process can be expected to happen over geologic time. Certainly if the problem of subsidence takes even hundreds of years, let alone thousands, it becomes relatively insignificant in human terms.

When subsidence is seen in human time frames, it is generally a localized problem directly over the mine. Essentially, it only attracts attention when the mine underlies structures that may fall into the sink hole caused at the surface by subsidence. Fortunately, the mined coal seams are commonly not very thick, and because of the rubble-packing process just described, it is quite unlikely for subsidence to be even as much as the height of the mine rooms. Nevertheless, a homeowner whose house falls even one foot will be seriously displeased. Mining companies find it necessary to take precautions to avoid any subsidence whenever they mine near areas of human habitation. Simply packing the mine with sufficient rubble or debris, though, generally controls this problem.

Mine Safety • Underground mines meet with generally less opposition than surface mines, although the hazards to workers' health are greatly increased below ground. The obvious factor that a person can only be buried in a cave-in if below ground should probably be overshadowed by the more insidious long-term threats to the miners' health. During the 20th century, more than 100,000 died in the mines, and another 100,000 retired miners suffered from Black Lung disease, as of a 1965 survey.[2]

The hazards of underground coal-mining faced by miners do not affect the environment or the population at large but take a serious toll on underground miners. Though these hazards are essentially impossible to eliminate, it is very possible to reduce them, and measures already employed have shown great progress. The use of steel or brick supports in combination with roof bolts and improved stress measurements and calculations have substantially decreased cave-ins. The incidence of fatal or disabling disease and injury are on the order of 100 per million worker hours in underground mines, which represents a 60% decrease from the hazards to underground miners before the implementation of the Mine Safety Act of 1969 and its 1977 amendments.[3] This demonstrates an impressive improvement in mine safety over the earlier part of this century. Technologies to improve ventilation so that harmful vapors and dust are not so concentrated and wearing filtration masks moderate the peril to the respiratory systems of men working below ground to provide energy. Improved ventilation standards are largely responsible for a significant reduction in mine explosions and fires as well. In spite of safety gains, underground coal mining remains one of the most hazardous of civilian occupations. Hazards to workers include the physical risks of cave-ins and explosions, as well as long-term health hazards from breathing coal-contaminated air. Even excluding violent, accidental deaths in the mines, statistics show that death rates for miners between the ages of 20 and 24 are 23% higher than the national average for male workers of that age. The death rate for miners aged 60 to 64 are 122% above the national average for their age group.[4]

Flooding presented a risk of disaster in early underground mines. If the miners broke through into a prolific aquifer, the mine could be suddenly inundated, drowning miners or cutting off their egress. A particularly dramatic tragedy in England in 1815 trapped 75 miners underground. They were robbing pillars in an old mine when water broke into the mine. It was months before the water was adequately drained for the mine to be reentered when all 75 were found in an unflooded chamber. They had starved to death over the months in subterranean blackness; some had only recently died. Modern pumping equipment is capable of handling huge volumes of water, though, even from great depth, greatly reducing the risk of flooding disasters in mines employing modern technology.[5]

Explosions of methane gas seeping from the coal seam into the confines of the mine were very common before the turn of the 20th century. Methane is explosive in concentrations of only 5% to 15% in air. In the early days of mining, open torches provided lighting, and an ignition source for accumulating gasses. The explosions could easily kill entire mining crews and start fires that could burn for years. Even as early as 1813, flame lamps were designed to separate the flame from the dangerous gasses of the mine. The earliest type employed a water seal and used a bellows to pump air through the water to the flame. While this technology provided for an effective seal and considerable safety, operating the bellows to introduce air was problematic for the miners who were otherwise occupied. Shortly thereafter, a wire mantle lantern was developed. The wire mesh conducted heat rapidly enough to prevent ignition of the mine gasses as long as the lantern was operated properly. To ensure this, the lamp was normally locked together for operation, and the miner had to take the lamp to a designated safe place where keys were kept if he needed to get into the lamp to make adjustments. The development of the battery-operated electric lantern offered the next significant step in mine safety, totally obviating the necessity of a flame to produce light. The electric lantern proliferated rapidly in the early years of the 20th century, with impetus provided by the Mine Safety Board.[6]

Before safety lamps were able to reduce explosion and fire disasters significantly, miners discovered that canaries were very susceptible to asphyxiation from hydrocarbon gasses. Thus, many caged canaries were carried into the mines to offer their lives as early warning signals of accumulating gasses and possibly provide the miners with adequate time to escape.

When a mine fire occurs, it depletes the oxygen in the mine, leaving the air abnormally rich in carbon dioxide and nitrogen. This air is heavier than oxygen-rich air, so it sinks to the mine floor. It smothers both flames and humans, and is called *black-damp* or *chokedamp*.[7] This can make escape and rescue efforts difficult and hazardous.

In addition to eliminating open flames, ventilation has been responsible for dramatic reductions in underground disasters. Ventilation has also been instrumental in controlling the incidence of respiratory diseases among underground miners. The most infamous disorder, generally referred to as *black lung disease* because the coal dust blackens the lung tissue similarly to cigarette smoke, is actually a variety of distinct disorders. Emphysema is common, as are several versions of pneumoconiosis. Perhaps the most pernicious of the diseases is one in which the ultrafine rock particles entrained with the coal are raised as dust to enter the respiratory system and wreak havoc. The fine particles can enter the small air sacks in the lungs and alveoli, irritating the membrane linings. Leukocytes (white blood cells whose role is to attack invading organisms) surround the irritation, but there is no invading organism to kill and break down. Some of the finest particles are so small as to be able to enter cells and slice genes in much the same manner as asbestos. These particles also can result in cancer.

The body is a resilient system and can commonly recover from such onslaughts if they are not excessively large or persistent. Unfortunately, the particles are so fine that no ordinary filter can effectively remove them. Keeping the mine rubble moist is one way to minimize the dust in the air that the miners must breathe. Safety measures like this have made for great strides in making underground mining a career more than a death sentence. Nevertheless, miner hazards are significant and should be borne as a cost of coal in any comparison of resources.

Transport • There are several impacts associated with transporting coal to consumption sites. The fatalities attributed to coal transportation are slightly higher than those caused in underground mines. Additionally, a good deal of energy is consumed in the transport phase of the massive solid. As discussed in the previous chapter, coal slurry pipelines offer a sizable savings in the energy lost in the transportation phase, but the large water requirements have raised strenuous objections to their implementation. Another indirect impact of coal use is on the development of mass transportation in the United States. Currently, since passenger services provided nationally by Amtrak use the same rail lines as freight services, the proliferation of slow-moving 100-car unit trains could increase the congestion of the rail arteries, choking off the potential appeal that mass transportation can offer. Furthermore, the fatalities attributable to hauling coal are even larger than from accidents in underground mines.[8]

Air Pollution • Probably the most pervasive problem with coal use is that of airborne emissions during combustion. The earliest efforts to address the massive emissions of smoke from large industries (including electric generating plants) were directed towards mitigating the sometimes severe effects on the local human population. In the latter part of the 19th and early 20th centuries, London endured disastrous accumulations of smoke from the growing industry and residential heating sectors, both fueled by coal. The heavy smoke aggravated respiratory disorders, especially among the elderly and infants. As many as 4,000 deaths were ascribed to the massive, concentrated air pollution in these incidents.[9] Of course, these numbers are virtually impossible to confirm since many of the victims had preexisting breathing problems, some of which might have flared up and taken some toll anyway. Even when no such dramatically intense smoke accumulations were experienced, the heavily populated, industrialized cities suffered impaired qualities of life from the smoke and smog. It is very true that these smoke emissions were related to coal combustion, but to blame the pollution on the introduction of the fossil fuel may be specious because wood fires providing the same amount of energy would have been responsible for similar amounts of smoke.

Since smoke is emitted at elevated temperatures due to the combustion from which it was produced, it is less dense than air and tends to rise into the atmosphere where it disperses and is carried away by winds. Accumulation of the pollutants, such as those that produced the killer smogs, requires unusual conditions referred to as *air inversions*. An inversion occurs when a body of cold air overrides a warm air mass. This is made possible in part by the fact that carbon dioxide is a more dense gas, at constant temperature, than oxygen. Therefore, as *blackdamp* lies on the floor of mines after fires, the deoxygenated combustion products can lie in the still air near the ground when weather conditions prohibit the normal rise of hot gasses. The first solution to eliminate (or minimize) the local effects of air pollution was to build taller smokestacks and chimneys. If smoke is contained until it rises sufficiently high, the wind disperses it broadly. It cannot fall back through cooler air to the ground to make a killer fog, nor can it remain onsite long enough to diffuse and mix with the surrounding air in significant concentrations. Smokestack height helps to dilute the atmospheric pollution but can do nothing to reduce its total volume.

Acid Rain and High Sulfur Coals • Only in the latter half of the 20th century did concern arise regarding the regional problems of air pollution. In the 1960s, first Europeans and then Americans directed concern towards acid rain effects from coal combustion. The sulfur which coals contain in widely varying concentrations is principally responsible for acid rain. Sulfur undergoes reactions similar to the carbon compounds, commonly yielding sulfur dioxide, just as the organic fuels produce carbon dioxide. As sulfur dioxide reacts with water vapor, it produces sulfuric acid. Carbon dioxide can also react to form some carbonic acid. The acid products are liquid at atmospheric temperature and pressure, thus they condense along with water vapors, literally causing a slightly acidic rain. This can be a very serious regional environmental impact, as has been seen very clearly in parts of eastern Europe where a great deal of high sulfur coal was burned with relatively little abatement for decades. Sufficient quantities of acid have been introduced to the forests and lakes downwind from the large coal-burning plants to lower the soil and water pH dangerously.

On the one hand, many scientists view acid rain as a manageable problem. It is clearly regional, not global, and acid-base reactions are chemically straightforward. One suggested means to restore the pH of acidified lakes is to dump limestone into them. The calcium carbonate in the limestone reacts with sulfuric acid to produce pH neutral salts of calcium and sulfur as well as water. However, this process results in increasing the water body's salinity and shifting the ionic balance of the water. Certainly, if dumping carbonate into an acidified water body is taken to an extreme, the increased *hardness* of the water can be as deadly to aquatic life as the acidity it fights. The tolerance of various species being exposed to salinity changes can be questioned, especially when already severely stressed by pH changes large enough to draw human attention and the decision to treat the water. Furthermore, the application of limestone to forest soils would certainly be more difficult to achieve and distribute uniformly.

A response to acid rain which is far superior to post-damage remediation is the abatement of acid emissions prior to disposal in the environment. Limestone scrubbers are routinely employed at the smokestacks of modern combustion facilities and can even be introduced during combustion in fluidized bed reactors. This technology holds much promise, and it may not be unreasonable to hope that production of acid rain from new facilities can be drastically curtailed with extensive dissemination of existing technologies. Technologies for particulate emission control fall into four general categories, in order of increasing effectiveness and generally increasing cost: mechanical collectors, electrostatic precipitators, wet gas scrubbers, and fabric filter baghouses. Mechanical collectors generally make use of the density difference between fine solids and gas. They may either simply offer an enlarged volume in which the particles can fall out of suspension or may employ inertial means to separate the solids, such as creating cyclonic flow in which the more dense material is accelerated to the outside of the cyclone. In electrostatic precipitators, the exhaust gasses flow between high voltage electrodes and collector plates. The electrodes discharge ions that strike the particles, charging them so that they are drawn to the oppositely charged collector plate. This process is impaired by the presence of substantial sulfur emissions, which tend to have high resistivity. The resistivity can be reduced by increasing the (already high) temperature of the emission gasses or by adding conductive material to lower the resistivity, which adds to the total emissions to handle. Wet gas scrubbers run the emissions through a water wash, which is very effective in removing contaminants but is costly to install. Finally, fabric filter baghouses (perhaps the most intuitively obvious technology similar to normal filtration methods) simply provide an enormous filter for the gasses to flow through which entraps the contaminants. Baghouses are actually very effective but costly, especially in terms of increasing the outflow pressure as needed to pass through the restrictions of the bag. They have found limited application in power utility plants because of this cost factor. Because these three most efficient options are costly, the mechanical (density) separators are often installed upstream to eliminate as much as they can and decrease the design load for the more expensive systems.[10]

Carbon Dioxide Emissions • The global effect of infrared absorption by *greenhouse gasses* such as carbon dioxide seems inevitable in the use of combustion energy sources. Coal produces more carbon dioxide per Btu delivered than oil and gas because its combustion and heat transfer are generally less efficient than that of fluid fuels. Converting coal to a liquid or gaseous fuel is an option that has received some attention. This process can be practical if access to naturally fluid combustion fuels is adequately constrained, and it does offer the environmental advantage of enabling precombustion sweetening (sulfur removal). However, the liquefaction or gasification processes produce emissions themselves, even though their products burn more cleanly. In addition, the well-tested Lurgi gasification process cannot use coal straight out of the mine with particles less than one-eighth inch in diameter. Some process to pelletize or agglomerate the small particles is required. The successful Lurgi projects are relatively small-scale, and though size is not a definite drawback, small-scale applications do not coincide with prevailing industrialized world energy-production trends.[11]

OIL AND GAS

Acquisition • Drilling and production activities create localized environmental impact. Onshore operations require clearing roadways and the drilling location itself, as do all forms of energy extraction technology. The size of the drilling site itself is very much a function of the depth of the well and scale of operations. It is very possible to drill to as much as 6,000 or 7,000 feet of depth with small, portable rigs, which along with ancillary equipment need occupy no more than the size of a city house lot. Deep wells not only require much larger drilling rigs but comparably larger mud pits, room for more and larger pumps, more personnel, and storage of materials to fight kicks and other down-hole hazards. However, even deep wells seldom require more than a few acres. Once a well is completed, most of the drilling site can be reclaimed in short order. In some regions, such as the United States' Appalachians, gas wells virtually disappear into the countryside, with the surface equipment occupying as little space as a single tree.

Offshore drilling has motivated a great deal of animosity, couched in environmental terms. Much of the drilling in the oceans utilizes platform structures which remain for the life of the field, and that can be several decades. In most cases, problems with the platforms represent aesthetic rather than environmental issues. The platform itself plants its steel feet on very small patches of the sea floor. Far from being destructive of marine life, the waters around the platforms often teem with life. Creatures such as barnacles attach themselves to the platform legs, attracting the creatures that feed on them, in turn attracting the creatures that feed on those creatures, and so on. One may ask if the creatures are being lured to death traps, and the answer would definitely seem to be, "no." The author has observed abundant marine life surrounding relatively old platforms in the Gulf of Mexico. Fishermen often try to get as close as they are allowed to harvest the abundant life, and when oil companies invite VIPs to visit their platforms they almost always advise, "bring your fishing rod." No evidence seems to have surfaced regarding toxic chemical contents in the shrimp or fish harvests taken in the vicinity of offshore oil fields.

Indeed, some of the oil companies facing the need to decommission platforms in depleted fields have proposed simply cutting the legs off and toppling the platforms into the water to form the seed for an artificial reef. These plans have met with approval, especially from the fishing industries who know that reefs make especially rich fishing grounds. Doubtless, this process does alter the local ecology. However, unlike the prompt flooding of lands to make hydroelectric reservoirs, the process is gradual and allows natural, endemic species to flourish over time. It would not seem likely that any species would have trouble relocating within the time frame of reef building. Nevertheless, it is very true that long-term environmental problems being recognized at the writing of this book were generally not foreseen. Detailed studies of the marine life at the reef, and for some miles around it, are doubtless appropriate. The evidence at the writing of this book, though, would seem to suggest strongly that there is no significant environmental damage associated with normal offshore drilling and production.

Leaks and Spills • The leakage of produced fluids during the commercial life of a field can contaminate the environment. Escaping gas adds to the *greenhouse effect,* and oil and formation water can contaminate surface waters and soil which create hazards to wildlife. Some natural gas inevitably escapes to the atmosphere during drilling and production operations as well as during transportation and consumption. Relative to the amount of fuel produced, the amount escaping is quite small; nevertheless, atmospheric studies show that methane concentrations have been increasing one to two percent per year during the latter part of the 20th century. A part of this increase can be attributed to leakage and venting of oil and gas production. Only about 10% of the total methane emissions to the atmosphere is estimated to be from natural gas releases. A significant portion of the increase seen in methane is due to increased ruminant livestock populations and rice agriculture.[12] Furthermore, a very large portion of the natural gas production that does escape to the atmosphere leaks from poorly designed and maintained lines in Eastern Europe. The application of the best technology has the potential to make pipeline leakages a very minor issue. In the long term, human exploitation of natural gas resources should ultimately diminish methane emissions. The gas is natural and natural tectonic activities create fissures and fractures which allow gas to vent to the surface. Those gas vents often serve as initial exploratory evidence of hydrocarbon accumulations. Once gas is produced, it can no longer escape to the atmosphere. The rate of venting from human exploitation activities is certainly greater than the background venting rate. However, after the era of gas use reaches its zenith in the coming century, human-induced emissions will drop, and the background emission rate will probably be lower than the original background rate. This argument seems to be somewhat important, albeit rather esoteric in the context of any single human lifetime, because the known environmental problem associated with gas is as a contributor to infrared absorption, 20 times stronger than carbon dioxide. In the sense that the greenhouse effect is a very long-term problem, the long-term decrease of methane emissions may not be trivial. Gas has no other significant known environmental effects.

The incidental release of water coproduced with oil and gas is also not likely to have dramatic environmental impact. The connate water (in which the rock sediments were originally deposited) most often is seawater. Unless the connate water has leached unusual, toxic salts out of the sediment, it should be no more harmful to the environment than spilling modern seawater. Unfortunately, the likelihood of leaching additional salts over time, or of the salts becoming more concentrated in some other manner, is not negligible. Some oil field brines contain very high salt concentrations and/or concentrations of toxic salts such as lead, arsenic, and boron. Considerable care should be exercised to minimize the entry of such toxic contaminants into the surface and near-surface groundwater systems. Of course, even ordinary seawater contains more salt than is potable. If it is released in sufficient quantities, it can increase the salinity of potable water sources beyond desirable levels. Most often, any formation water produced in significant amounts is gathered and reinjected into the original formation or some other zone identified as not containing or being in pressure communication with potable waters. Many times, in oil fields utilizing waterfloods for secondary recovery purposes, the reinjection of produced water is a useful part of the operation. However, waterflood fields ultimately produce and reinject a great deal of water, some of which inevitably leaks.

Probably the most serious and without a doubt the most visible pollution from oilfield operations is the spillage of oil. Modest amounts of oil leak from surface valves, transfer, and storage equipment under normal operations. While the cumulative leakage across the decades-long life of oil fields is substantial, the operations of well-maintained fields probably do not create significant environmental damage. Oil seeps, like gas vents, have occurred naturally for millions of years, and nature has its own responses to and uses for crude oil. Notably, there are microbes that digest crude, turning it into biotic material. (The process of digestion is as opposed to ingestion, in which an animal takes crude into its digestive system but has no means to break it down into food chemicals.) Of course, the crude also breaks down in natural oxidation reactions on the earth's surface. Petroleum is not particularly poisonous to most animals, including humans. This fact can be attested by the dubious but apparently harmless practice of using petroleum as a medicine before the advent of the petroleum industry. It is when oil enters the environment in large volumes within a short range of space and time that damage is done. Most of the oil spills that have attracted attention of people born after 1950 have been results of the transportation process rather than drilling or production. Indeed, in 1985 the National Academy of Sciences conducted an extensive study of the *Origins, Fates, and Effects of Oil in the Oceans.* The study found that offshore drilling and production operations contributed relatively minor oil contamination to the seas. The study also found that offshore operations have even more minor impact because most of the oil contamination occurring in these operations is in small amounts, drawn out over long time periods. The exception to this finding is oil-well blowouts. When a well blows out, the formation fluid escaping from the well often consists primarily of water and gas, but an offshore blowout from an oil-bearing zone has the potential of creating a large oil spill.

PEMEX suffered one of the most spectacular and tragic offshore blowouts in the Gulf of Mexico in 1980. An estimated 3.3 million barrels of oil were lost as the well blew out of control for several months.[13] This was unquestionably an environmental disaster. Modern drilling and blowout prevention technologies greatly reduce the likelihood of such accidents. In the United States during the period from 1971 to 1989, 82 out of 19,450 offshore drilling wells experienced blowouts. Most of these wells did not spill any crude oil, and only a total of 70 barrels of crude were recorded lost to the environment in all 82 wells. (Many of the blowouts occurred in gas or water-bearing zones which produced no oil.) During the same time period, another 44 wells blew out during completions, production, or remedial operations, spilling a total of 830 barrels—a volume much more significant than the amount spilled in drilling operations but still minuscule in comparison to the amounts spilled even in minor tanker accidents.)[14]

Employment of the best technologies can and does minimize the risk of blowouts and other significant accidental discharges of oil from offshore operations. While the risk cannot be reduced to zero, it is already substantially lower in drilling and production than in shipping oil.

Shipping Spills • The Exxon Valdez tanker, coincidentally leaving the port of Valdez, Alaska with a load of crude oil from the Alyeska Pipeline, ran aground in Prince William sound on March 24, 1989, discharging an estimated 259,000 barrels of its 1,260,000-barrel cargo into the ocean. The impact was tragic. The slick spread and covered 10,000 square miles of near coastal waters, and the mortality toll on marine mammals and birds was huge.[15] Workers found 1,000 dead otters and 30,000 dead birds. (While it can be argued that the total mortality was even larger, workers acknowledged that some of the animals recovered may have died of natural causes.)[16] The disaster received massive media attention, as well as a cleanup response effort in which Exxon deployed more than 3,000 people, and the Coast Guard more than 1,000.[17] The policy ramifications of the furor raised by the Valdez disaster remain unclear but will be discussed in the final chapter.

While the Valdez accident was tragic, it should not be allowed to obscure the reality that tanker spills occur with some regularity; there are hundreds of spills every year. This accounts for approximately 30% of the total oil entry into the oceans each year.[18] Almost all spills incur serious environmental costs because they are concentrated beyond the tolerance and capacities of the local environments. This is particularly true in the case of marine and inland waterway shipping. Oil floats on top of the water and moves away from cleanup equipment, and wave action carries it over containment buoys. The floating oil fouls the fur and feathers of aquatic mammals and birds that unwittingly swim or dive into it.

Longer-term effects of oil spills include the contamination of shellfish which ingest oil as they strain seawater for food. The oil they ingest but cannot digest is likely to be stored in their tissues to be ingested by predators who consume them and cannot digest the oil either. Since the oil has relatively low toxicity, it must be ingested in rather large quantities by the shellfish to kill them, but it can make their flesh foul-tasting in much smaller concentrations. This can impair the commercial potential of oyster beds and perhaps disturb the local food chain.

The aesthetic impact of oil spills is driven sharply home to the television viewing audience. The oil that washes onto shore blackens beaches and forms thick, tarry masses as the lighter components evaporate. The visual impact may be illustrative of environmental damage but is not synonymous with it.

Cleanup measures usually start with containment. Oil naturally tends to spread rapidly across the water surface into a relatively thin layer. Once it has spread, recollection of the oil and removal from the environment generally costs time and money. So, booms are typically deployed around the oil slick. Booms are buoyant flattened plastic tubes which float partially submerged. The oil floating on top of the water cannot pass the boom unless wave action carries it over the top of the containment. This, in fact, represents the most serious limitation of boom containment. Experts have observed that only 50% of the time are wave conditions likely to permit successful use of booms. When they do work, they can greatly limit the damage and allow remedial efforts to focus on a limited body of water.

Physically removing the oil from the ocean is the most desirable step. Skimming it from the surface can be very effective. Some specially designed skimmer ships have flat bows that lower to just below the waterline. As the ship proceeds ahead slowly, the uppermost fluids are skimmed into the ship's interior, where they can be pumped into the ship's hold. Of course, these vessels cannot operate effectively in rough seas. By their very design, they also take in a great deal of water in addition to the oil. It takes many skimmer vessel loads to gather any significant portion of the cargo of a Very Large Crude Carrier.

As the oil slick spreads and is carried to shallow near-shore waters, it can be mopped up from the surface of the water. Materials such as cellophane and straw have strong oil-wetting tendencies, which is to say that oil adsorbs readily to their surfaces. A vast human labor pool can be well employed in this stage of cleanup. The mopping process is very much like mopping a kitchen floor. It should not be difficult to imagine that each man or woman mopping vigorously would be doing very well to gather a barrel or two of oil in a day. Straw can be spread on the slick to pick up much of the oil. Then the oil-soaked straw can be gathered with pitchforks. In addition to wading on the shore and in the shallows, human workers can stand on barges if the water is calm.

Large portions of the oil slick are likely not to approach the coast in shallow or calm enough water for mopping operations. The spreading slick that has eluded containment and recollection remains hazardous to wildlife, roughly in proportion to its thickness and concentration. Thus, dispersal, the opposite of containment, may become advantageous. Surfactants, which reduce the interfacial tension between oil and water, like those used in oilfield tertiary recovery (or in kitchen detergents), may be applied to the slick. This allows the oil to become soluble in the water, greatly speeding the natural dispersion process. Of course, rather than removing the oil contaminant, this adds another foreign chemical to the environment. Nevertheless, the slick presents an extreme environmental hazard and one that calls for drastic action. Furthermore, once the oil is dissolved, the oceans do offer an enormous sink to dilute and return the oil to biotically safe forms.

Another extreme remedial step is to burn the slick off the water. This can eliminate the oil quickly and minimize the hazard to wildlife, as creatures are much less likely to swim or fly into fire than into an innocuous-looking slick. The crude oil burns with a great deal of smoke. In shallow waters, the heat may not be able to disperse broadly enough and could conceivably take a serious toll on aquatic life, particularly of slow-moving creatures such as mollusks.

One of the important aspects of spill response is definitely an effort to stem wildlife mortalities. This is an effort that has probably been witnessed by more people than any other; its pathos conveys readily in the media. Few can view without emotion the struggles and deaths of beautiful marine creatures. Any acceptance of the sanctity of life places an ethical burden on humans to extend care and attempt to mitigate the death and suffering imposed on innocent lives by human activity. Even from a purely utilitarian view, some of the creatures most imperiled by spills are endangered species, whose loss threatens the richness of the global gene pool. Others are creatures who have commercial value. For the mobile surface fishing birds and mammals, preventing their entry into the slick could be the best remedy. Warning devices to scare them away can be somewhat effective; however, the senior biologist on site at the Alyeska cleanup indicated that the deployment of devices to scare away birds was decidedly futile, speculating that it may have had more political or emotional motivation than based on reality.[19] For the many animals that do become oil-fouled, attempting to catch the animal and clean its fur or feathers is the best one can do. Sea mammals can die of hypothermia as the crude dissolves their body oils and mats their fur that keeps cold water from their skin. Birds suffer similar problems as their feathers become matted with oil and useless for warmth or flight, allowing them to become oil and water-logged and sink. In spite of the large response of animal-care personnel, thousands of animals can be expected to die in a large spill. As some of the dense portions of oil sink to the bottom to be ingested by bottom-dwellers, there is little that can be done for them.

The best recourse for oil spills is to avoid them. Since by far the greatest risks of releasing dangerous amounts of oil to the marine environment comes from shipping, energy self-sufficiency does become an environmental as well as a national security or economic issue. Reducing the amounts of oil moving across the oceans in tankers is perhaps one of the best ways to minimize the oil spills. Additionally, using pipelines whenever possible reduces the risk greatly. Finally, the design and safe operations of the tankers that must ply the waters is essential.

Vehicular Fuel Demand • Fluid fuels are capable of more efficient combustion and heat transfer under many circumstances than solid fuels, but the character of consumer demand is equally important. When Edwin Drake drilled the first oil well, the fuel was intended for lighting to replace dwindling oil from over-hunted whales. Thus, oils that yielded a great deal of kerosene upon distillation were preferred. Heating and lighting can be reasonably efficient uses of petroleum fuels, and the demands in heating, lighting, and cooking sectors could be supplied by oil and gas for many years. The development and proliferation of internal combustion powered vehicles turned the demand profile around dramatically, as well as creating a huge increase in demand.

Unfortunately, the unprecedented mobility and related benefits offered by the automobile came at a very large energy cost. The efficiency of converting chemical fuel energy to motive work in the engine is barely 15%. This is less than one-half the net efficiencies of industrial heat processes and even of typical electric consumption in the home. Furthermore, the final work output of the vehicle in moving passengers or freight is even smaller. Thus, the nature of the demand for transportation service dictates a high subsequent level of consumption per unit of work delivered and a high level of environmental impact.

For every mile driven, the average American car puts into the atmosphere nearly a pound of carbon dioxide in addition to other pollutants including oxides of nitrogen, carbon monoxide, and particulates. Estimates indicate that 90% of toxic carbon monoxide, 80% of benzene, and half of gaseous hydrocarbons come from cars and trucks. This is, in part, the price of a modern, mobile society. It is also a function of consumer preferences. A car with one passenger is responsible for practically four times as much pollution as a car with four passengers on a per task basis. A car with a large, powerful engine, capable of rapid acceleration is approximately 20% less efficient than a comparable vehicle with a smaller engine. The choice to use public transportation systems could use as little as one-eighth as much energy as private cars.

The demand for private vehicle transport in the United States has ancillary environmental impacts that should be considered along with the additional impacts of land use and contamination associated with mining, drilling, and shipping energy. In the United States, approximately 33,400 square miles of land are taken up by highways, not to mention parking lots and other pavement for vehicles.[20] Millions of animals are killed annually on those highways—animals whose death and suffering are not covered by the television networks. Similarly, vehicular transport accounts for nearly 50,000 human deaths annually in the United States alone. These numbers easily exceed the real and probabilistic mortalities associated with acquisition and processing of all energy forms.[21] These impacts are not a function of the consumption of energy per se, but of the mode in which it is consumed. Consider, for example, the individual who chooses to drive one-half of a mile rather than walk that distance. Total consumption of energy for this task is increased by a factor of 100, and pollution by a larger factor because of the imperfect high temperature combustion in the engine. The time required to walk that distance is approximately 10 minutes at a fairly leisurely 3 mph pace, while the drive would take about 3 minutes (generously assuming that the car needs only to be entered, started, and driven at an average 10 mph over that distance, without stopping for traffic or pedestrians). The driver also statistically incurs a likelihood of killing someone or being killed in the trip, on the order of 1 in 100 million. Although this is not a significant risk in itself, if put in the context of such a discretionary vehicular use being selected 100 times during a given year, it amounts to a risk similar to living next door to a nuclear power plant. Of course, it can be argued that most fatalities occur at higher speeds than our hypothetical driver achieves, so it is more likely simply to cause an injury; indeed, it is about 100 times more likely.

KEROGEN

The processes for extracting and retorting kerogen are normally simultaneous (in the case of in-situ retorting) or continuous (in the case of mining and onsite retorting). The mining does require the extraction of massive quantities of rock, since the synthetic crude extracted is likely to be only a few percent (by weight) of the shale. The UNOCAL Corp. shale oil project that proceeded to demonstration scale employed bench-type mining in the cliffs of northwestern Colorado. Unlike conventional strip mining, the bench mine enters the formation at an outcropping. The mining proceeds in a fashion similar to room and pillar underground mining, with large pillars of rock left to support the overburden. Very little vegetation was disturbed in this commercial-scale synthetic crude venture. The greatest environmental issue raised with the use of oil shale stems from the volumes of waste fine rock material generated. Since the shale is crushed and retorted, it is essentially all reduced to extremely small particle sizes; however, since the total rock volume increases after retorting and synthetic crude extraction, it is impossible to replace all of the debris in the excavation. The fineness of the particles permits them to become easily wind-borne. Keeping the tailings piles wet can ameliorate this problem but requires vast quantities of water. Potentially toxic salts can be leached from the wet tailings, ultimately entering the hydrologic cycle and contaminating surface and subsurface sources. Perhaps variations of recent technologies to fuse ash into paving material could provide a long-term solution for the disposal of spent shale debris. The question appears to be moot in the foreseeable future, as the commerciality seems dubious even when energy prices rise again.

BIOMASS

The direct combustion of biomass has virtually no redeeming values in terms of environmental impact. Regions in which people consume primarily firewood and charcoal for energy experience extreme loss of forest (even if the energy only serves minimal domestic needs). This alone seems an adequate response to the notion of a *closed-carbon-cycle.* A closed cycle refers to a system in which the cycled stuff continues to cycle and recycle without accumulation or loss. The closed-carbon-cycle then, refers to a system in which carbon released from plant matter by oxidizing it to atmospheric carbon dioxide is completely recycled back into the newly growing plant matter, without any accumulation or loss of carbon from either side of the cycle. Clearly, this cannot be the case, where one step in the supposedly cyclic process destroys the recycling capacity of the system. As each tree is cut down in a deforested region, there is less biotic capacity to take up and fix the carbon which is released in combustion. In other words, carbon is lost from the biomass side of the cycle and accumulates as carbon dioxide in the atmospheric side—by definition, an open cycle.

In a system with full replacement of biomass—where the rate of regrowth equals or exceeds the rate of consumption—a closed carbon cycle may well describe biomass burning. The real system, modern-day planet earth, does not even closely approximate the requirements for a closed system. Indeed, technically, it does not make any difference whether firewood and charcoal consumption are major contributors to forest loss. As long as forest loss is the situation, any consumption that does not contribute directly to reafforestation adds to atmospheric carbon dioxide buildup and to deforestation.

Additionally, the direct combustion of firewood and charcoal is much more polluting on an energy equivalent or per task basis than most "fossil fuels," with the possible exception of coal. While London's "killer smogs" were the result of smoke produced by burning coal, the total combustion energy produced in that city had grown dramatically from the time before the broad switch from firewood to coal. In 1600, the population was roughly estimated at 300,000;[22] by 1900, it had grown to about 2 million. Even more important than the increasing energy demand to meet basic human needs was the Industrial Revolution. To gain some perspective on the difference, consider that prior to the Industrial Revolution it has been estimated that the average energy consumption per capita had grown very slowly from roughly 40 Btu/person/ day, to just under 90 by the advent of industrialization. By the late 19th century, when the first killer smogs were recorded, the per capita energy consumption had increased to over 260 Btu/person/day.[23] The proliferation of automobiles and electrification within the next century tripled consumption levels again. With ten times as many people consuming three times as much energy per capita, it is not unreasonable to believe that wood-burning at such a level would have produced a similar disaster, if it had been feasible to sustain such consumption levels. In fact, some evidence of this can be drawn from the 17th century writing of Mr. John Evelyn. He observed that, "a hellish and dismal cloud" settled over London, such that "catarrhs, consumption, and coughs rage more in this one city than in the whole earth besides."[24]

Even at the preindustrial levels of consumption, England's demands greatly outstripped forest growth rates, and deforestation was a major factor in the progress of coal use. It is quite true that lumber demands for ship and homebuilding were instrumental in deforesting the British Isles, but firewood demands added to the trend. Reducing the stress caused by the energy sector of wood demand was important in stemming the tide of forest loss. Actually, some authors dispute the deforestation of England in the years before the major transition to coal. Some even argue that the firewood production from forests covering as little as 2% of England's land surface could easily have met the country's energy requirements.[25] The firsthand accounts of deforestation cited by many authors is more persuasive than those authors who hypothesize that deforestation could not have occurred. (Similarly, the passengers of the Titanic observing firsthand the icy waters of the North Atlantic engulfing their "unsinkable ship" would carry more credence than all of the experts who theorized that sinking was not possible.) One error that those arguing against the observed deforestation of England might be making is to ignore the numerous additional purposes wood from the forests served, such as in construction.

It is often argued that if fuelwood is provided by special fuelwood plantations, where new trees are planted for each tree harvested, then the cycle is closed and fuelwood demands can promote restoration of forests planted as fuelwood plantations. Clearly, though, the global carbon cycle remains open. Furthermore, the production of carbon is by no means the only environmental impact of any energy source, especially biomass. Even cursory examination of the Short Rotation Intensive Culture firewood plantation can dispel any notion of such applications restoring forest in any ecological terms. It is clear that the plantings are far too closely spaced and quickly harvested ever to be considered trees, or certainly to provide a habitat for wildlife. These plantations have no more apparent potential for preserving the environment than any other form of modern, intensive, mechanized agriculture.

Biomass in Nonindustrialized Countries • A biomass production deficit is seen clearly by many Third World residents, who depend upon energy from firewood and charcoal. Some western researchers claim that there is no real fuelwood shortage. Such claims seem to fly in the face of reality and often contain basic logical errors. In a discussion of developing alternate energy resources, e.g., oil and gas, the retired Dean of the University of Ghana rose and said, "Once you could not see for the forests, now you can see as far as you want."[26] Again, while energy demands are not the only and perhaps not even the dominant cause of forest loss, they unquestionably contribute to it.

The consumption of biomass fuels in the Third World is typically inefficient. The technologies to facilitate dramatic improvements in the efficiency of conversion and consumption exist and can moderate the damage to some extent. Difficulties in disseminating such technologies effectively prevail and will be evaluated in the final chapter. Hours in the smoky cooking enclosures present a health hazard to many Third World women. Little research seems to exist regarding the level of these hazards, although African government agencies report observations of increased incidence of pulmonary cancers amongst fuelwood-dependent women. The absence of data does not rule out this impact.

Studies also indicate that the development of firewood plantations will incur substantial worker hazards, even in affluent western settings, where there is abundant, enforced, safety legislation. The studies expect disabling injuries to exceed the rate of such injuries in underground coal mines. Interviews with missionary doctors indicate a prevalence of firewood cutting and gathering injuries in fuelwood-dependent regions.

To some extent, biomass harvesting which is restricted to gathering already dead material evades the arguments just presented. If no plant life cycles are terminated for the sake of energy production, then the use of biomass does not directly contribute to deforestation or environmental degradation. Even this exception, though, has a limit. Consider the example of harvesting agricultural waste in the form of stubble remaining in a corn field. Certainly that stubble is already dead and can no longer serve as a sink to take up carbon dioxide. If left, the stubble decays in the field. Therefore, harvesting it as a fuel feedstock would clearly make sense, at least superficially. However, even if plant material is dead, it continues to serve a role in the local ecology. The corn stubble provides some protection to the ground against wind and rain erosion during the fallow season. It also provides nutrients to the myriad of small creatures dwelling in the soil. Their burrowing and excrement are nature's tilling and fertilizers. The less organic debris is left on the fields, the more energy intensive and chemical intensive the agricultural system must become.

Methane from Biomass • Anaerobic digestors which convert decaying biomass principally into methane gas present an entirely different set of environmental factors than does the burning of firewood or charcoal. Anaerobic digestors offer a means of deriving useful energy from true waste, which is otherwise destined either for use as an agricultural fertilizer or for disposal in sanitation systems. Since the residue from an anaerobic digestor has excellent fertilizer properties, there is no obvious environmental cost to using excrement for the production of energy in this manner, other than the emissions of combustion. (The nonobvious cost may be taking nutrients away from creatures that live in the soil.

That seems to be an argument to be resolved between advocates of modern, chemical-based and organic agriculture.)

Making use of the chemical energy in human excrement seems not to have any environmental cost, other than emissions. Under the pressures of modern population levels, it is highly desirable to gather and treat sewage, which has no other practical use. Large-scale sewage treatment plants in the industrialized world have collected and made use of the methane gas naturally generated by microorganisms in the excrement. The methane provides most of the power requirements of the treatment facilities but little else. Since the emissions are the only environmental impact associated with bio-gas, the net impact can actually be beneficial. Decaying organic matter, such as excrement, naturally emits methane. If the methane simply evolves into the atmosphere, it contributes 20 times more to the greenhouse effect than it does as carbon dioxide.

Ethanol Production • The concept of producing alcohols from biomass has some mass appeal because it can replace oil in the high-demand sector of vehicular transport. Relatively toxic aldehydes are generated as incomplete combustion products of ethanol in internal combustion engines, as well as comparable amounts of other emissions except carbon monoxide. Furthermore, corn and sugarcane, which offer some of the best feedstock biomass for alcohol production, are normally grown in highly energy-intensive agriculture, which limits the overall efficiency of the fuel produced. Sugarcane, too, is notorious among the historically problematic crops in terms of environmental disruption. One factor in the United States which might mitigate the impacts related to corn production is the surplus corn supply maintained by the U.S. government. If nonenergy social factors make the stockpiling of corn cost-effective, then using older portions of the stockpile for fuel production, to be replenished by fresh supplies, can be argued not to place the environmental cost of growing the corn on the fuel. Environmentally, though, ethanol production appears to be of marginal or even questionable benefit in relation to the fossil fuels currently in use. Its advantage would fall more reasonably under the category of renewability, which is decidedly a separate issue from environmental benefit.

Energy from Waste • The utilization of the chemical energy content of human refuse also avoids the caveats regarding harvesting living biomass for energy. The landfill space not used by refuse burned to produce heat is an environmental benefit accruing to this energy source. The ash occupies a small fraction of the volume occupied by the original waste. Currently, refuse-derived combustion fuels are limited to waste streams exclusive of hazardous chemical wastes. The production of toxic chemicals during combustion is still an issue, though. Dioxins, poisons that are deliberately produced for use in herbicides, can also be produced during the combustion of some plastics. However, if the wastes are burned at temperatures greater than 1,800°F for more than one second, they can be broken down to harmless chemicals.[27]

However, the topic of hazardous waste incineration for useful heat recovery receives some attention. While it can release some toxic gasses to the atmosphere, it eliminates the need to bury the waste in a specially secured site, which will inevitably breach at some time in the future, disgorging the toxins to the environment. While the time of breach is likely to be far in the future in human terms, many chemical toxins, unlike radioactive waste, can be expected to remain as potently toxic centuries hence as at their time of burial. The toxicity removed from the subsurface environment is a benefit of incineration. The production of useful heat is also a benefit to be compared to the net impacts of existing energy alternatives. This will be considered in the final chapter.

NONCOMBUSTION SOURCES

SOLAR POWER

Solar energy seems to be the golden child of many environmentalists. In many applications, it appears to have tremendous advantages over most combustion resources. However, some impact is unavoidable. Land use is an issue sometimes raised in terms of the environmental impact of solar energy resources. In fact, under the efficiency limitations of prevailing technology, active solar devices do require substantial collector areas. The fact that most solar collectors are relatively stationary means that it is possible to place the devices on rooftops of vehicles as well as buildings. The limiting impact here is more likely to be one of aesthetics than of environmental damage. It must also be considered in comparison to the land use requirements of other energy sources.

Passive Solar • Land use, however, is not a factor at all for passive systems, as they necessarily incorporate the structure itself, and passive solar space heating is likely to utilize smaller buildings (with the possible exception of space for storage). Passive applications have the least impact associated with them. The collector is simply the house or appliance designed to make use of heat and light from the sun. Insofar as practical passive solar space heating requires minimizing losses, superinsulating a structure designed for solar heating may offer the sole potential source for negative environmental impact. The synthetic polymers from which the most effective insulating materials are made have been found to emit some toxic gases, including gaseous urea, aldehydes, etc. The total quantity emitted is very small for each structure and is prolonged over many years. In terms of polluting the global environment, the emissions can have very little effect, especially when compared to conventional combustion sources of heat. The environmental hazard is, in fact, localized to the interior of the specific structure. Newer solar house designs include some means of circulating air between the interior and exterior to prevent buildup of minute toxic gas emissions. A heat exchanger can minimize the loss of heat in the air circulating process which is generally deemed adequate to account for the only negative impact of direct passive solar use.

Solar Thermal Electric Conversion (STEC) and Indirect Heating • In these technologies, solar radiation heats a fluid which then carries thermal energy. In areas where the ambient temperature in which the system operates is likely to be below freezing, the fluid must either be selected to have a low freezing point or have anti-freeze chemicals added to a water medium. Either way, the chemicals employed do leak to the environment and create some damage. Furthermore, to prevent the growth of algae and fungi, biocides must also be added to the circulating medium. Since biocide literally means *life killer,* it is clear that it is toxic. For STEC plants, estimates suggest that as much as 23.6 liters of potentially contaminated water can be expected to be discharged for each gigajoule of electricity generated.[28] This is a non-negligible impact. If massive implementation of this technology were to proceed, the toxic discharges could be expected to amount to as much as 300,000 gallons of waste water per day for each gigawatt power plant.

Photovoltaic • While the actual energy production process is as pollution-free as one can imagine, toxic chemicals are commonly used in the manufacture of photovoltaic cells. The quantities are modest but then so are the power outputs of the best contemporary cells. For instance, arsenic is often used to *dope* the negative (*n*) side of a silicon photovoltaic cell. The arsenic atoms replace some of the silicon atoms in the crystal structure or lattice. Each arsenic atom provides an extra valence electron, adding to the conduction and producing an overall negative charge in that crystal. In order to create an electrical junction, the *n* silicon must be placed in contact with *p* (positively charged) silicon. This side of the junction can be doped with molecules such as boron, aluminum, or indium, each of which has one less valence electron. The doping is done in minute concentrations, on the order of one part arsenic per million parts silicon. Therefore, it would be reasonable to assume that even if photovoltaic cells were mass-produced to provide a large share of the production of electricity, they would not present any serious toxic waste disposal problem at the ends of their useful lives (and of course all devices do have finite productive lives). It seems possible to envision some risk of spillage in the manufacture of the cell when large plants are producing hundreds of millions of them. A calculation of the risk-discounted expected value of contamination should be possible, based on the total amounts of toxic material to be handled in large-scale factories and the incidence of chemical spillage in other similar manufacturing operations. Doubtless, the environmental hazard to be expected is not very large, especially when compared to other energy sources.

WIND POWER

Wind power produces no chemical pollution in its own right. As with any intermittent power source, the storage system may be a potential source of negative environmental impact. Wind farms do require a significant amount of land. It can be argued that the land between wind turbines can be used for other purposes, such as agriculture. In addition, modern turbines produce substantial noise pollution. The classification of noise as a pollutant, thereby having environmental and/or health impacts, seems open to debate. The evidence does seem to suggest at least some impact on the mental or emotional health and well-being of people subjected to constant noise, even if the noise is not loud enough to jeopardize the person's auditory system. The resolution of the impact on human and nonhuman life by the noise of wind turbines is beyond this author's scope. In view of the growing awareness that the unseen or sensed effects of electromagnetic radiation from high voltage power lines may be linked to cancer, the issue of noise having physical effects on living creatures does seem worthy of consideration.

There is a definite aesthetic issue associated with the proliferation of wind farms. They do require siting in prominent places, such as hilltops and plateaus which might well have scenic value. Of course, some may find the towers with whirling arms pleasing, as this author finds the stark silhouette of an offshore platform on the horizon. Many, though, can object to the landscape being cluttered with artificial structures well away from towns and cities.

The rapidly whirling blades of the horizontal axis turbines do conjure an image of a potential, localized hazard. Experience with large-scale machinery suggests that structural failures do occur when millions of hours of work are logged. Consequently, large-scale wind farms are probably best not located in areas which share use with humans, just as oil field pump jacks are commonly fenced to prevent people from being accidentally injured by the large equipment. Hazards to birds have already been documented. If livestock shares the space, some losses can probably be expected, but based on the market values normally assigned to nonhuman and nonendangered species, the costs are likely to be too small to calculate.

None of the preceding problems can be assigned to the slower moving, old fashioned windmills that have dotted the European landscape for centuries. (Granted, the occasional individual may find that even these offend some aesthetic values.) Where relatively slow and intermittent rotational work is required, as in grinding, the old-style windmill should invoke virtually no negative impact.

One impact that may be common to all of the tall wind-gathering structures is the possibility of interfering with radio waves. Some complaints have been recorded to this effect. Siting and the already established requirements for remoteness should keep this issue minimal.

Similarly, it is difficult to imagine a negative environmental, health, or aesthetic impact associated with a return to some increase in wind-powered marine shipping. Clearly, the statistics cited earlier in this chapter from the National Academy of Sciences show that even beyond the savings of combustion fuels, sailing vessels could incur significant decreases in oil contamination in the waterways, including harbors. The added cost in terms of losses in time value of the shipments to the greater time required for sailing vessels seems to be the sole and overriding negative impact associated with this option. The time value costs may be offset by the adoption of sailing vessels which employ backup diesel propulsion.

HYDROPOWER

Water power was once viewed as the near-perfect option for energy production. It creates no chemical pollution whatsoever and is relatively efficient. The secondary effects of ecologic disruption, though, have drawn considerable fire and a dramatic reversal in views toward this energy source. Centuries of small-scale use did not prepare anyone for the level of impact to be realized by modern applications.

Disruption of Local Ecology • The most dramatic impacts are the direct result of flooding large tracts of land to create a water reservoir behind a dam. A large lake is suddenly created where a river was once bounded by land likely to be a fertile flood plain. Often the dam is placed in a canyon whose walls retain the water depth required to turn the large turbines. In the case of deep, narrow canyons, perhaps the least immediate environmental damage is incurred. In these cases, there are no fertile flood plains in the canyon, though upstream lands also flood as the water level rises to the dam height.

In any case, the flora and fauna of the area are suited to river and dry land ecologies, not lacustrine. Those land animals not sufficiently mobile to move drown. The makeup of the fish population normally changes substantially from various species of trout and salmon to dominant lake fish such as bass, bluegill, and perch. Some of the smaller fish varieties may be displaced altogether such as the well-publicized case of the snail darter—a small fish facing extinction that halted the construction of a major American dam. Trees and other plants are immobile and have no chance to move. While geologic processes constantly change the shape of the earth's surface, raising parts of the seabed to mountain tops and submerging forests and plains, these changes take many millennia. The changes wrought by dam building happen far too quickly for adaptation. They also are most often conducted on large rivers far beyond the cyclic changes generated by natural dam builders like beavers.

Changing Patterns • A more pervasive and less localized set of phenomena relate to the less conspicuous changes caused by the dam project. The migration routes of spawning fish may be disrupted. Even by the middle of the 20h century this problem was recognized, and dams built in the Pacific Northwest incorporated *fish ladders* to help salmon negotiate around the huge artificial obstacle. Their success seemed apparent as even passers-by could stand at a railing during spawning season and watch the salmon leaping spectacularly up the steps of the ladder. However, biologists have begun to raise the warning flag that these clever alternatives are not working nearly as well as had seemed to be the case. Apparently not all of the fish find the ladder, and some exhaust too much of the energy stored up for the arduous trip on which they do not eat. Spawns are seen to be falling off over the years since these dams were erected.

In addition to the fish it supports, the river itself is disrupted by the dam. Rivers are massive natural transport systems carrying millions of tons per year of sediment to the oceans. The velocity of the flowing water allows it to carry those sediments. During natural flood years, the increased volumes of water erode more rock and pick up more sediment to carry and deposit on the flood plains which make them very fertile. Dams disrupt this process. The huge volumes of sediment are dropped when the water slows to a virtual stop in the reservoir. The result is that the reservoir silts up with sediment and curtails the useful life of the dam. It also disrupts downstream depositional patterns on which some other ecologic systems have developed. Dams can even cause a recreational disaster by cutting off sand being carried to maintain natural beaches.

Recreation is the value added to offset the economic loss to farmers and the aesthetic loss to tourists. Studies indicate, though, that the recreational opportunities are somewhat exclusive and are enjoyed by modest numbers of people who can afford boats and recreational trips. Even if the exclusivity were mitigated by policies to facilitate use by inner city dwellers, the substitution of a recreational benefit for an environmental cost is specious.

As with wind, the historic uses do not create the sort of negative impacts associated with modern uses, largely because of the huge difference in scale. The old watermills could be situated on the banks of a stream, using its natural though modest force. A mill race, commonly built to divert a portion of a stream through a sharper elevation drop, might kill some unfortunate aquatic creatures but did not disrupt the entire local ecosystem. The need, though, for such small-scale energy applications in modern times seems very limited.

Tidal and Wave Power • Tidal power operates on similar principles to hydroelectric dam power but may have less of an associated impact. If the sluice arrangement is built in a portion of the tidal flat, the flora and fauna already there are adapted to periods of inundation. Some tidal feeding animals may be caught in the turbine inlets, but a high incidence of this does not seem likely.

Wave power generators show no clear negative impacts. Some may become apparent if technology and economics ever make them a viable energy source. The transmission of electricity generated offshore may be a source of problems, especially if the mutagenic potential of corona effects is established.

GEOTHERMAL POWER

The use of geothermal energy to produce electricity is unusual amongst the energy sources that do not use combustion, in that it can have airborne emissions in the case of open-loop systems (systems which run steam produced from the formation directly through the turbines, such that some can escape to the atmosphere). The formation waters in the rocks exposed to abnormally high temperatures from greater depths may also contain toxic salts, such as those of boron and arsenic. The release, then, of produced water on the surface in shallow aquifers has the potential of creating serious damage to freshwater sources. These risks can be minimized by operating with a closed-loop system, in which the produced water drawn from the wells flows through a heat exchanger where it transfers much of its thermal energy to a second fluid, which in turn runs through the turbine, while the geothermal brine is reinjected into the reservoir. In this case, the potentially polluting formation water can only escape through a system leak, whereas in the open-loop system, some of the steam inevitably escapes to the atmosphere through the turbine outlet. The closed-loop system also protects the turbine from the corrosive, saline waters, but it is impossible to transfer all of the heat to the carrier fluid. This problem is offset by the fact that a more volatile carrier fluid that vaporizes can be used, which can drive the turbine at a lower temperature. Of course, some of the thermodynamically preferred fluids, such as fluorocarbons and ammonia, have their own environmental costs, and their losses must be minimized as well. The total contamination per unit of energy produced is minimal for properly managed geothermal projects.

Ocean Thermal Electric Conversion (OTEC) • Some question is raised about the potential for environmental impact from the mixing of deep, cold waters and warm shallow waters at OTEC sites. This mixing has the potential to cause marine animals to move up or down out of their normal depth strata. More likely, it may cause some animals to move away from the altered ocean thermal gradient. The upwelling of nutrients from the ocean depths may also cause a sudden growth of plankton. This effect would probably be quite localized. But some scientists believe that the alteration of the thermal profile in the oceans could cause carbon dioxide to be released from the water. Even this problem should not produce a serious environmental cost insofar as the estimates suggests that if carbon dioxide is released, it would still only amount to one-third as much of the gas as is released in conventional combustion-fueled power plants.[29]

The carrier fluids circulated may include fluorocarbons or ammonia, which have the potential to leak and create toxic contamination. This energy source does not seem to be developing to a point that clarifies the impacts significantly.

NUCLEAR POWER

Just as solar and wind seem to be the darlings of many environmentalists, nuclear fission energy is often lumped together with fossil fuels as one of the villains. Radioactivity can be dangerous when highly concentrated—but what can't? Much of the publicized debate about nuclear energy contains emotional reactions mixed randomly with logical arguments. The use of nuclear energy definitely presents a set of risks, which may or may not overshadow the risks and costs of the combustion fuels with which it practically competes.

MYTHS

The debate is cluttered with half-truths and implicit assumptions that need to be clarified and made explicit. Is radioactivity unnatural? It certainly is not. Probably every student reaching secondary school learns about the man-made trans-uranium elements, which are radioactive and typically have very short half-lives. These elements have short half-lives because the overcrowded nuclei are so unstable that they quickly undergo radioactive decay. Perhaps that has created the notion of nuclear radioactivity as artificial. These artificially created elements are very rare, however, and insignificant contributors to radiation outside of the laboratory or reactor in which they are made. The radiation released under normal operations of large nuclear power plants increases the level of radiation received by those living in the plant's immediate vicinity by approximately 1%.[30]

Everyone on the face of the earth encounters radiation every day, on the order of 100 millirems per year. (A millirem is one thousandth of a rem. A rem is the amount of ionizing radiation that produces the same effect on a gram of living matter as one *rad* of radiation from x-rays. A *rad* refers to the amount of energy imparted to a gram of living matter from radiation. The distinction is that the amount of ionization is a better indicator of the health effects of radiation than simply the energy it imparts to tissue.) This is the natural background radiation level—it was present long before the Curies conducted their pioneering experiments in nuclear radioactivity. The discussion of lithology formation evaluation in Chapter 2 brought out the fact that the naturally occurring radioactive isotope of potassium is prevalent enough, to make natural gamma radiation in the clays large enough to distinguish them from nonpotassium-bearing rocks on the basis of radioactivity. Clay is the principal ingredient in brick. This accounts for the statement that a brick house emits more radiation than a normally operating power plant. And it is true.

Is radioactivity more dangerous than other forms of pollution? Occasionally wild stories can be heard of sites that are so radioactive that "if you were to drive past it at 60 miles per hour, you'd get a lethal dose of radiation." That would be incredibly radioactive. Radioactivity and its effects are probabilistic: an intense dose of radiation delivered equally to a roomful of people would kill some, but not all of them. The median lethal dose refers to the dose required to kill one-half of the people exposed to it (within a specified time period). For humans, the dose is estimated to be 400 roentgens.[31] To put it in terms of the unit of radiation used previously (rems), an exposure of the whole body to a brief dose of 100 rems causes no immediate effects. A dose of 450 rems would be the median lethal dose, while 1,000 rems ensures death, and 8,000 rems produces prompt incapacitation. Even from the blast of a 1-megaton nuclear warhead, the median lethal dose extends only for about a 1-mile radius.[32]

The three basic kinds of radiation have very different levels of damage. Their potential to kill descends through the Greek alphabet: alpha is the most dangerous, then beta, and gamma radiation. Fortunately, the tendency to interact with matter which makes them dangerous also makes their ability to penetrate matter inversely proportional to their dangerousness. A sheet of paper can stop a majority of alpha particles, a thin sheet of metal excludes beta particles, but a foot of concrete is required to stop gamma rays. It is clear, then, that if the driver rolls up his or her car windows while driving past the supposed deadly site, virtually no alpha particles will enter the car and very few beta particles, but the gamma ray dose may be high if there is no containment of the vast radioactive stockpile. On the other hand, if the source of radiation is buried a few feet below the surface and covered with cement, very little radiation should be leaking. However, the site may still be dangerous because the containment materials inevitably become somewhat radioactive as they absorb the energy and mass of radiation and ultimately become emitters themselves. So, the radioactive material stockpile site might not be a good place to stop for a picnic and certainly not to build a house. The capacity of even large quantities of intensely radioactive material leaking from containment to kill someone in a brief exposure, exemplified by driving by at high speed, is extremely limited.

Perhaps the notion of the extraordinary deadliness of radioactivity results from the devastating near-term and long-term effects of nuclear weapons. The confusion of nuclear power with nuclear war merits clarification. Within 1 mile of the site of a 1-megaton atmospheric detonation virtually everyone is killed and essentially all structures are leveled. The immediate thermal radiation (heat) is sufficient to kill most of these people. Within seconds, people outside are also exposed to a shock wave capable of killing them.[33] The prompt radioactivity can be lethal in this range as well, but even if there were none, the people in the immediate blast have no real chance of survival. This is an entirely different phenomenon than the most spectacular failures conceivable for nuclear power plants. For one thing, the megaton weapons use fusion reactions to produce the extreme explosions. Fission bombs, using uncontrolled versions of the same reaction as nuclear power plants, are much smaller. In the event of a sudden catastrophic failure of a nuclear power plant, the workers remaining inside the plant can be at very high risk, but most of the workers can be evacuated. Residents within a few miles should also be evacuated in such a case. Since it is not an explosive release, people have much more time to respond than to nuclear warfare. After the final containment failure, all of the plant's releases will be carried along no faster than winds. No residents will simply drop in their tracks (unless their fear of nuclear radiation induces a heart attack).

Isn't nuclear radiation especially dangerous because of the long half-lives of some radioactive isotopes? The fact is that many chemical toxins never lose their toxicity at all. The radioactive decay phenomenon which produces half-lives is an advantage that nuclear power wastes have over chemical toxins. For instance, the DDT once applied liberally to fields, forests, and neighborhoods remain in the environment to threaten generations to come. Anyone who clings to the myth that chemical toxins are not cursed with the longevity or potential danger of nuclear materials may want to read Rachel Carson's *Silent Spring*, which certainly does not advocate proliferation of nuclear power but does provide a uniquely compelling picture of chemical hazards in the environment.

HAZARDS OF NUCLEAR POWER

Nuclear power does have serious environmental issues with which to contend. Waste disposal is one. In every year of operation, a one-gigawatt nuclear power plant produces about 5,000 gallons of liquid, high-level radioactive wastes, and much larger volumes of low-level radioactive waste. The half-lives seem to bring attention to the length of time during which the material remains dangerous, and some experts have named the waste disposal as the most serious impact of nuclear power plant operation. The waste does remain dangerous for many years, and leakage to freshwater supplies could be especially disastrous. Some of the policy issues related to handling the wastes will be discussed in the final chapter. The sobering reality is that the total number of lethal doses of radioactive waste is grossly exceeded by the lethal doses of chemical toxins, which are used and not disposed of in any controlled manner. The annual production of fly ash and sulfur dioxide by conventional coal-fired plants is more than 1,000 times larger than the volume of radioactive waste produced by a nuclear power plant in generating the same amount of electricity.[34]

Catastrophic failure is an issue brought home dramatically by the actual failure of the one gigawatt Chernobyl power plant in the Soviet Ukraine, if not by the near failure of the Three-Mile Island reactor near Harrisburg, Pennsylvania. The design of the Chernobyl unit precluded a containment vessel, which is standard on most western units. Furthermore, the Chernobyl design allowed the reactor core to reach a *prompt critical* stage. In this type of reactor, the neutrons directly released are sufficient to sustain a chain reaction. In other reactor designs, the neutrons emitted from delayed, subsequent reactions are required to make the reaction critical. In the latter design, as reactivity increases, there is time to insert the control rods before the delayed reactions. At Chernobyl, this safety factor did not exist. Several operator errors during a low power test were added to the design flaws. One error included shutting off the emergency cooling system to facilitate measurements in the test. The reaction went out of control, melting down the reactor core, starting fires in the plant, and releasing 50 megacuries of radiation to the atmosphere. The short-term fatalities of this catastrophic failure amounted to 31 plant personnel and firefighters, and an additional 168 people suffered acute radiation sickness. The long-term impact is estimated at an additional 10,000 to 40,000 cancer deaths in Europe and the former Soviet Union over the subsequent 50 years. These numbers are tragic but reflect less than a .015% increase in the expected number of cancer deaths for that region and time period.[35]

The Chernobyl disaster clearly mandated designs that incorporate containment structures and nonprompt-critical operation, which should offer great potential for decreasing the likelihood of a comparable disaster. It would seem to be some vindication of the prevalent U.S. reactor design that more than 20 operator and equipment failures were documented in the Three-Mile Island incident, without catastrophic failure.

Even the workers in nuclear power plants are monitored to ensure that the radiation to which they are exposed is a small fraction of the dosage that is considered dangerous. Workers in the uranium mines experience risk similar to those of underground coal miners. However, a ton of uranium ore produces about 100 times more energy than a ton of coal. So, to provide equal amounts of energy, the nuclear power industry exposes far fewer miners to the risks than does the coal industry.[36]

Weapons proliferation is also cited as an impact of nuclear power plant proliferation. It is possible. There is no such thing as perfect security, so the chance of a terrorist group stealing weapons-grade fuel does indeed exist. Of course, and perhaps more easily, a terrorist group can mine uranium, build a small enrichment plant, and make their own. The other issue of proliferation is the possibility that a nuclear reactor in a country with an extremist, expansionist, or otherwise nasty government, can build a nuclear power plant and use normal operations to generate weapons-grade materials. Of course, this is possible too. Unfortunately, the proliferation of nuclear weapons technology is the reality. It would seem, nuclear power or not, that a government that really wants nuclear weapons can get them. Consider the transactions of nuclear devices that occurred and nearly occurred between the United States and Iraq while hostilities were imminent during the Persian Gulf crisis of 1990.

The reality is that experience has exploded the myth of limitless, clean, safe energy to supply a utopian society by fission reactions. Nuclear energy is the most intense source known and thus the most difficult to control. Nuclear accidents do have a unique potential for endangering the lives of people for miles around the plant who do not work in the energy production industry. Yet meeting the energy demands of industrialized world citizens costs lives and burdens the environment. Available data do not show fission-based nuclear power to be clearly worse than other commercial sources.

Fusion power has the potential to be cleaner and safer but is totally unproven as a controllable reaction. One of the costs which must be assigned to nuclear fusion, as to other unproven energy sources, is the expense of research to develop the technology. It can certainly be argued that the billions of dollars spent by governments on nuclear power research (especially controlled fusion) could provide immense benefits if applied to the development in other areas, such as the redevelopment of extensive, reliable mass transportation infrastructure in the U.S., energy efficiency technologies, or even of solar or wind technologies. Expenditures in research are speculative. While experts can assign expected probabilities to developments from various kinds of research, only the end results of the research can determine which technologies truly have commercial-scale potential.

ENDNOTES

1. Mandelbaum, Paulette 1984, *Acid Rain*, Plenum Press, NY.
2. Braithwaite, John 1985, *To Punish or Persuade: Enforcement of Coal Mine Safety*, SUNY Press, Albany, NY, p. 1.
3. *Energy in Transition 1985-2010: Final Report of the Committee on Nuclear and Alternative Energy Systems 1979*, National Academy of Sciences, W.H. Freeman and Co., San Francisco, p.196.
4. Braithwaite, *To Punish or Persuade*, 1985, p.1.
5. Ashton and Sykes, 1964, *The Coal Industry of the Eighteenth Century*, Manchester University Press, Manchester, pp. 40,41.
6. Moore, Elwood S., 1940, *Coal*, Wiley and Sones, NY, pp. 305-307.
7. *Dictionary of Mining, Mineral and Related Terms* 1968, United States Bureau of Mines, Washington, D.C., p. 108.
8. Scientists' Institute for Public Information, 1975, p. 257.
9. Fisherman, Jack and Kalish, Robert 1990, *Global Alert: The Ozone Pollution Crisis*, Plenum Press, NY, pp. 199-205.
10. *Clean Use of Coal*, 1987, OECD, Paris, pp. 105-115.
11. *Synthetic Liquid Fuels*, 1985, OECD, Paris, pp. 90-100.
12. Smil, Vaclav 1987, *Energy, Food, Environment*, Clarendon Press, London, p. 257.
13. *Collier's Yearbook 1990*, MacMillan Educational Services, NY, p. 111.
14. *Basic Petroleum Data Book 1991*, American Petroleum Institute, Section XI, Table 14.
15. *Oil in the Seas: Inputs, Fates and Effects*, 1985, National Acedemy Press, Washington, D.C., p. 35.

16. Davidson, Art 1990, *In the Wake of the Exxon Valdez*, Sierra Club Books, San Francisco, p. 131.
17. *Collier's Yearbook* 1990, pp. 110–111.
18. Davidson, *In the Wake of the Exxon Valdez*, p. 186.
19. ibid.
20. *Transport and the Environment*, p. 35.
21. Kraushaar, Jack J. and Ristinen, Robert A. 1993, *Energy and Problems of a Technical Society*, 2nd Edition, Wiley and Sons, NY, p. 329.
22. Elton, G.R. 1969, *England Under the Tudors*, Metheun and Co., London, p. 229.
23. Cook, Earl 1975, *Man, Energy, Society*, W.H. Freeman Co., San Francisco.
24. Fisherman and Kalish, *Global Alert: The Ozone Pollution Crisis.*
25. Buxton, Neil K. 1978, *The Economic Development of the British Coal Industry*, Bratsford Academic, London, pp. 19–20.
26. Dr. Richard, Orocco-Tete, personal communication at the 1993 African Studies Conference in Accra, Ghana.
27. Wollard, Katherine 1988, "Garbage In, Power Out," *IEEE Spectrum*, 25: pp. 42–44.
28. *Energy in Transition*, pp. 371, 372.
29. Kraushaar and Ristinen, *Energy and Problems*, p. 138.
30. Kraushaar and Ristinen, *Energy and Problems*, pp. 443, 444.
31. *A Dictionary of Mining, Mineral and Related Terms*, p. 636.
32. Kraushaar and Ristinen, *Energy and Problems*, pp. 440-445.
33. ibid, pp. 440–445.
34. Walker, Charles A., et al. 1983, *Too Hot To Handle*, Yale University Press, New Haven, CT, pp. 71, 72.
35. ibid, pp. 128–130.
36. *Energy in Transition*, p. 216

CHAPTER 6

ENERGY CHOICES

With an understanding of the basic technical factors determining the availability, production, and end-use of energy resources, along with the direct costs and indirect costs associated with each step of energy acquisition and utilization, it is possible to begin assessing energy alternatives to meet needs in the present and the near future. By design, this chapter raises more questions than it answers. Some unusual perspectives are presented to help readers step back from and question "received wisdom." Errors are inherent in replicating wisdom that may have been valid in contexts that dictated different sets of assumptions, that may have seemed valid under more limited constraints on information, or that may even still be valid.

There is no single answer to "What is the best energy resource?" Each resource has been shown to have a broad range of costs and benefits in its acquisition, transportation, storage, conversion, and end-use. The context and applications of energy use are important. Solar power plants may become viable in arid lower latitudes but will probably never be practical in Iceland where geothermal energy has long been used for space heating. It is possible that residents of regions facing drought and famine caused by deforestation would be very willing to take the risk of mining coal underground. Although this book cannot provide many definitive or sweeping answers about energy options, it is possible to develop an approach to evaluating the options. An effective approach includes all of the costs and benefits from the whole range of activities involved in tapping each energy resource. The final analysis requires assigning relative values to the costs and benefits of each energy option in each context, and comparing the summation of values to assess viable energy mixes for each region. This book is meant to offer a starting point for approaching such a comprehensive analysis of energy systems.

ABUNDANCE AND AVAILABILITY

Abundance is distinct from environmental impact, though discussions are often carried on as if the issues of abundance and of renewability were identical with issues of the environment. Concern over the long-term availability of energy resources can be legitimately focused on efforts to include renewable or nondepletable sources into the energy mix. The potential to deplete the reserves of an energy source is an environmental issue only when those reserves are part of the ecosystem. This would be true for biomass but not for the fossil fuels.

SUPPLY AND DEMAND ISSUES

Demand level is closely tied to supply and price levels in neoclassical economic theory. When supply exceeds demand, prices tend to fall as sellers try to increase their market share. This tends to increase the level of demand as consumers find that they may increase the benefits they derive from the good or service by consuming more without paying more. Theoretically, prices continue to fall for an oversupplied good until an equilibrium is established in which supply roughly equals demand. As prices fall, some supplies become cost-ineffective and are forced out of the marketplace; this also facilitates reaching a new supply/demand equilibrium. Conversely, if demand exceeds supply, the price increases. The sellers can increase their profits by raising their prices, without losing market share, until a new equilibrium is reached. In this scenario, demand decreases as people can afford less of the product. Supplies which were not cost-effective at a lower price equilibrium enter the marketplace. If supplies do not reach demand, prices continue to skyrocket. Sooner or later, the prices become so high that switching to other kinds of supplies becomes worthwhile to consumers (e.g., switching from firewood to coal in deforested 17th and 18th century Europe or from petroleum to one of the alternative sources at some time in the future). This is the basic law of supply and demand. One important caveat is that, while economists with perfect access to supply information may be able to calculate ideal supply/demand equilibrium curves, consumers' perception of supplies of goods they deem essential is more important than real supply numbers. This was demonstrated clearly in the response to perceived supply shortages in the 1970s: panic buying, skyrocketing prices, and eventually conservation.

An additional complication relative to energy supply/demand estimations results from the peculiarities of oil and gas production. The largest production costs are encountered in developing reserves, while producing the reserves has minimal unit cost (especially for oil and gas). Therefore, the owner of hydrocarbon reserves has limited incentive to decrease the production from those reserves.

Although coal dominates fossil fuel reserves, oil and gas dominate the market place and thus determine energy prices. If there were no OPEC, and no efforts to constrain production, the owners of the supergiant oilfields could conceivably increase their production to the point of flooding the market and driving the price of oil to a point so low that most American and European production would become unprofitable. A large portion of American production comes from enhanced recovery (including secondary) and from *stripper* fields in which each well produces less than 10 barrels of oil per day. Furthermore, a large portion of European production comes from offshore in the hostile waters of the North Sea. All of this production is more expensive than onshore wells on primary production. Exploration would virtually cease because sufficiently depressed prices would not allow the high costs of modern exploration to be recovered in an acceptably short time. However, even the supergiant oil fields are definitely being depleted. When they have been depleted to the point that they are being produced at or near their capacities, prices will rise to match replacement costs. If a great deal of oil remains to be discovered, new discoveries will replace the declining production and prices will stabilize at higher levels. When the time comes that discovery rates falter in spite of the technological advances in prospecting, however, reserve additions from new discoveries will not replace reserve drawdowns, and prices will continue to increase until other energy sources become cost-effective.

RESOURCES AND RENEWABILITY

The recoverable resource base for coal under existing economic and technical constraints is larger than for any other fossil fuel. Although biomass receives attention for its apparent renewability, every resource has some rate of renewal function. The depletion of resources relates not only to their renewal rates but to the sizes of the reserves from which they are drawn. While the renewal rate for coal is clearly slower than that of biomass, the stock from which it can be drawn is much larger. For instance, if the United States were to replace the nominal 20 quadrillion Btus produced by coal with firewood, it would require 2.3 billion tons of wood each year. (This figure was achieved using similar calculations to those shown in the following discussions of Britain's transition from wood fuel to coal.) Therefore, if less than 2 million square miles of natural forest were used to supply this need, it would exceed renewal rates. That represents more than half of the total area of the nation and far more than the area of the nation's natural forests.

But the preceding *reductio ad absurdem* argument is not quite fair. Many of the proposals for the development of biomass fuels to meet modern demands focus on the derivation of secondary fuels from vegetable crops. These can include alcohols derived from the fermentation of agricultural wastes, including corn stalks, sawdust, etc. It can also include pressing natural oils from crops such as soybeans. These plant oils can be fairly readily refined for use as diesel fuels. In terms of renewability, soybean farms can be developed to provide an indefinitely renewable yield to offset a portion of coal, nuclear, imported oil, or offshore oil use. The fuel's renewal, though, must not be confused with environmental benefits. The environmental impact of converting more arable land to monoculture farming is highly suspect, as described in the preceding chapter.

SUSTAINABLE FIREWOOD PRODUCTION

The ultimate sustainability of fuelwood use is a myth that richly deserves an ignominious death. Biomass is a unique energy source. It certainly is potentially renewable in the human time frame, but its renewal rate is based on the size of its reserve base. The more forests there are, the more wood grows. Therefore, as people consume firewood at rates that exceed the local forest's growth rate, that consumption activity decreases the amount of wood that will grow in the future.

While some seem to view firewood as categorically renewable, and some researchers even argue for the concept that firewood consumption for iron-making in England was indefinitely sustainable, this does not stand up to analytical scrutiny. By the end of the 19th century, 200 million tons of British coal supplied 4 quadrillion Btus of raw chemical energy every year.[1] Let's assume a forest with growth rates at the high end because natural growth can be at 8 cubic meters per hectare per year. This equates to about 114 cubic feet/acre/year or 3,700 pounds/acre/year (taking an intermediate wood density.)[2] If the green wood contains about 4,300 Btu/pound chemical energy and the coal contains a typical 10,260 Btu/pound, the sustainable production of an equal amount of chemical energy from firewood would require harvesting the total growth equivalent of about 265 million acres every year. This translates to more than 400,000 square miles. The fact that the entire area of the British Isles is less than 150,000 square miles presents a problem that becomes immediately evident. Firewood alone could not have fueled the growing industry and population of England.

It can be argued, though, that the 8 cubic meters growth rate per hectare per year is very conservative. The biomass production of Short Rotation Intensive Culture plantations described in Chapter 1 can be as high as 50 cubic meters per hectare per year.[3] So, with a factor of 6 increase in total productivity, the Intensive Culture (*wood grass*) system would be capable of providing England's late 19th century energy demand if only one-half of England, Scotland, Wales, and Ireland were planted with wood grass. Of course, the environmental impact of creating such extensive monoculture would not be positive. Besides, this does not consider inarable lands, lands occupied by humans, or the land required to grow food. This example shows biomass to be unsustainable as the main energy provider for population levels well below those of modern society with only basic industrialization.

In order for biomass use to be sustainable, the scenario must be changed in one or more fundamental ways.

1. Biomass must be made a much smaller contributor to the formerly fuelwood-dependent society's total energy mix.

2. Technologic improvements can permit the useful production of work from less input energy.

3. The biomass can be produced from multiple use plantations.

All three of the above options are probably significant in order to use biomass effectively. (Hypothetically, lowering the population level or the standard of living provided by industrialization would also enhance the sustainability of biomass energy production, but neither is practically or socially viable.)

The first option is fairly clear and seems imperative. If biomass is a relatively small contributor to modern energy production, it can be indefinitely sustainable, and at the same time, it can ease the depletion pressure on fossil fuel resources. Unfortunately, in a world facing the combined threats of massive deforestation and atmospheric disruption from combustion fuel burning, the role it can play without creating environmental damage must be questioned. Direct combustion has been shown to be at least comparable to fossil fuels in terms of carbon dioxide and particulate emissions.

Producing energy more efficiently is an objective for every energy source. Numerous projects have been underway to introduce more efficient wood-burning stoves in fuelwood-dependent regions. It is true that the efficiency of cooking over an open wood fire is extraordinarily low (on the order of 10%), and that stoves designed to control the airflow can achieve efficiencies exceeding 30 or even 40% in laboratory tests. However, actual field experience shows that practical efficiencies show very small improvements over open fires.[4] Several problems exist, such as the size and shape of wood available as described in the conversion and end-use chapter. If logs do not fit inside a stove for the stove's door to be closed, neither the air flow nor the efficiency is controlled. A means of addressing this problem is converting biomass to liquid or gaseous fuels that permit the use of very efficient appliances. It has been noted as well that compressing wood chips into pellets permits the use of stoves that burn not only very efficiently, but so hot that very little smoke or particulate emissions result. Unfortunately, the pelletizing process is very costly. On the other hand, the processed fuels have the advantage of burning more cleanly. This is especially true of methane derived from biomass.

The option of multiple use can address the land requirement problem. Land need not be devoted solely to the production of fuel, and firewood does not need to be the only use of the plants harvested. To the extent that crops already being grown for food can be used simultaneously for energy production, the biomass does not need to take on the entire ecologic impact burden of taking land over for monoculture production. In some cases, biomass can be produced from waste byproducts of other processes, such as use of the sawdust from lumber operations. Finally, combined-use agriculture, such as the planting of tree rows along the edges of fields, can provide small farmers with firewood from the prunings. If the trees also provide fruit or nuts, the multiple-use criterion is especially well honored, and the tree roots can hold soil to prevent erosion.

THE FUTURE OF OIL AND GAS PRODUCTION

About two-thirds of known oil resources are not recoverable under prevailing technologic and economic constraints, thus another very large source of reserve additions exists. Enhanced recovery techniques have already been developed which would permit recovery of a large part of the previously unrecoverable portion of the resource base.

Dramatic increases in price can be expected when production capacity or deliveries are interrupted. Furthermore, since exploration is a probabilistic venture, with significant lead time prior to commercial production, dramatic market speculation is probably inevitable in times of short supply. Much of the increase in energy prices during the 1970s was a response to this speculation. Such speculation can almost always be expected to overrun the real shortage; it occurs among merchants in the marketplace and cannot adequately account for new discoveries being made. Those new discoveries will not find their way onto the companies' book reserves until extensive confirmation drilling and substantial production proves their magnitudes. This fact is dictated by the definition of reserves, which are proven to be producible under existing technology and economics. It would seem quite predictable that the next energy supply shortage is likely to reflect previous supply shortages. Prices will rise and reserves will increase. Companies already know where substantial oil and gas resources lie. They are not reserves because they are not shown to be producible under current economic constraints, but a sufficient increase in price would instantly make some known resources become reserves.

In terms of availability, none of the fossil fuel resources are particularly close to "running out," although the exploration and recovery costs for new reserves to offset production will ultimately increase. At the time of this writing, the existing export production capacity of crude oil in major producing nations exceeds global demand. Therefore, reserve replacements are not required in the market place, and the price of production from existing reserves does not reflect the increasing cost of replacement. Inevitably, as long as the existing productive capacity meets or exceeds demand, owners of reserves will seek to increase their profits by maximizing their production, which keeps prices low.

COAL: THE BACKSTOP ENERGY RESOURCE

Coal reserves are far more abundant than the reserves of any other combustion fuel. Since combustion can meet some forms of energy demand more effectively than any other primary energy source, and since the fluid hydrocarbons appear to have limited potential to increase their global market share in the coming decades, coal will most likely be a significant energy resource for generations to come.

Even if noncombustion energy sources take over the marketplace, coal is likely to remain a significant commercial product. It is composed primarily of carbon, which is the building block of organic compounds, and can generate an incredible array of synthetic products. Currently, oil and gas are the primary feedstocks for synthesizing organic compounds, but it seems very reasonable that even after the vast bulk of the potentially recoverable oil and gas resource bases are depleted, coal production may take over the synthetic products market niche. The niche is small relative to energy consumption (on the order of 10% of total oil consumption). However, in the absence of cheap oil and gas, it may be a very important and lucrative future market.

THE NONDEPLETABLE SOURCES

Solar, wind, and hydro energy are referred to in this book as *nondepletable* sources because human-use does not deplete these resources. Human technology has the potential to tap into these energy flows but not to alter their resource bases. If human activities were to alter regional climate patterns, it is conceivable that changing rain patterns could affect river flows and thus hydropower potential, but if anthropogenic changes reach that level, humanity probably has worse problems than maintaining hydroelectricity.

All of these resources carry vast amounts of energy globally. Regional availability is the only constraint on the quantities of any of these resources. Some authors already suggest that the majority of high quality river sites have been exploited in Europe and the populous 48 states. The best sites for generating electricity from wind are generally not in the most desirable regions for habitation. Sunny regions are desirable, but society has developed such that a large share of the energy-consuming populations of industrialized nations are concentrated in the northern latitudes where sunshine is not ubiquitous. Nevertheless, none of these resources are constrained by their availability: it is other costs.

Geothermal energy is lumped in with the nondepletable resources in this book, although it is possible to reduce its local abundance by over-use. Like the others in this category, though, it is not feasible for human activity to drain its source (cool the earth's core). It is constrained, like hydropower, by the location of its best sources but not ultimately by its abundance.

NUCLEAR POWER

The earth has a great deal of uranium, but most of it is so broadly dispersed as a trace constituent of rocks and ocean salts that it is not likely ever to add to the reserves. Indeed, if fission power is to grow in importance and be sustainable in the long term, it would appear that breeder reactor technologies will have to dominate. Breeder technology is available and probably cleaner than prevailing technologies. Breeder reactor deployment has doubtless suffered from public concerns about nuclear energy and the consequent moratorium on fission reactor development in the United States.

Fusion is constrained not by abundance but by availability. Some estimates suggest that there is enough deuterium in the oceans to provide energy to humanity almost as long as the sun will. But will technology ever make that vast energy resource available? Some technologic breakthroughs will be required for that.

ACQUISITION

Estimates of resource abundance for the fossil fuels are constrained by limits to exploration and by the technologies required to reach into the earth and extract the mineral resources. The same can be said of geothermal energy and uranium fuel for nuclear power. Finding solar, wind, biomass, and hydropower is not a limiting step; however, harvesting them can be limiting. The direct and indirect costs of tapping resource bases are very important in assessing the future energy production from each source.

COAL

Extractable deposits of coal are known to exist abundantly at relatively shallow depths in industrialized nations. Evidence regarding consumption patterns suggests that coal will not regain its 19th century market share due to environmental costs, lower end-use flexibility, and mining hazards. Although advances in mining technology have dramatically reduced the hazard to the life and health of coal miners in the 20th century, underground coal mining remains an inevitably hazardous occupation. Working hundreds of feet beneath the earth's surface necessarily exposes workers to the hazard of cave-ins. In evaluating safety measures, it is prudent to recognize that the roof supports must function in a dynamic rather than static system. The rocks are in constant motion, and the mining activity produces new sets of stresses in the rock. The deeper a mine extends below the surface, the risk becomes greater, and greater is the expense to moderate the risk.

Additions to coal reserves in the industrialized world can be expected to proceed in rather predictable fashion, while vast additions to the global base may be made from time to time as exploration and production commence in previously underexplored lower income countries. Coal resources have been identified in many countries that currently have limited or no production. The technology required to reach shallow coalbeds is very limited.

Therefore, firewood and charcoal shortages can be expected to drive increasing exploitation of the coal resources in these countries, much as the same shortages drove a transition to coal in England prior to the Industrial Revolution. The transition to coal and increased mining defines new reserves and holds the probability of uncovering vast new coal fields. Many existing efforts to mine coal in places such as Pakistan are conducted with technologies comparable to those employed in 17th century England. The mine shafts are small and inadequately supported. The same material (wood) whose shortage provides impetus to coal mining is the material that would provide timbers to shore up the mine passages. So, of course, there is a high incidence of cave-ins. Battery-operated lanterns are not readily available, so the miners use open lanterns as their predecessors in Europe and North America did a century ago. The risk of mine explosions is everpresent because of these open flames, as it was in earlier mines. A high incidence of respiratory disorders can be expected in the coming years due to the limited ventilation.

OIL AND GAS

Much speculation centered around the remaining productive life of fluid fossil fuel resources during the "oil crises" of the 1970s. The continental United States has received more than a century of vigorous exploration, with production established in more than a half million wells and a few million wells drilled. This exploration intensity has delineated the regional resource base. Significant new field discoveries are not likely in this well-explored region. However, more than two-thirds of the original oil in place remains in the reservoir at the end of a typical field's economic life. If the price of energy increases dramatically again, it becomes immediately commercially viable to continue producing fields about to be abandoned due to diminishing flow rates. It also will become viable to employ more of the sophisticated and expensive enhanced recovery technologies which were largely mothballed with falling oil prices in the early 1980s. Additionally, reexamining abandoned fields and old *dry holes* with modern geophysical and petrophysical techniques is likely to identify bypassed reserves. As the saying goes, "The best place to look for oil is in an oilfield."

Exploration on the European continent followed closely on the heels of the birth of the North American oil industry. Most of the exploration in Europe was very disappointing. Depending on where one chooses to draw continental boundaries, the early Russian industry and successful gas developments in Holland may have been the principal exceptions. A century later, offshore drilling technologies permitted the discovery and development of large oil fields in the British and Scandinavian North Seas. Since oil prices were on the downturn by the time some of these costly fields were reaching maximum production, exploratory drilling remains limited at the writing of this book. Significant new reserve additions can be expected if the oil price rises again to levels that make drilling in hostile (i.e., dangerous and difficult) waters profitable. Relative to global demands, though, the European continent seems to have limited potential to add to the reserve base.

The opportunity for new field discoveries to offset global reserves depletion may be quite high. Much of the world remains to be explored. For example, only 4,000 wells have been completed on the African continent, which encompasses six times more land mass than the United States, where more than half a million wells are producing oil and gas and literally millions have been drilled. Doubtless, many millions or billions of barrels of oil and trillions of cubic feet of gas remain to be discovered on this vast continent, as local or global demand drives exploration activities there. The same can be said of many South American and Asian regions. Indeed, many sizable fields have been discovered already in nonindustrialized countries in locations relatively remote to export facilities. These resources can be expected to enter the global reserves tally when the world market price rises sufficiently high to make development for the export markets attractive, or when development technologies are deployed within those countries for local use. Production for a local market can always be done less expensively than production for export. This is especially true for gas, since local pipelining can be very cost-effective, while liquefaction for export requires a large capital investment followed by significant operating expenses.

The popular comparison of reserves to annual production suggests a remaining productive life of petroleum reserves of less than 40 years. But an understanding of the meanings of these terms readily debunks the notion of running out of oil in that time period. A better estimate can be obtained by plotting reserves on a logarithmic scale against time on a linear scale. Such a plot (Fig. 6–1) suggests that nonreplaced reserve draw-down is on the order of 1% to 2% per year. If the prevailing trends of gradually declining replacements continues and petroleum demand remains more or less constant, the global reserves would fall below 10 times global demand around the year 2035. (The ability of production to meet demand will probably be exceeded by about the time that reserves are 10 times demand level, as it is not possible to produce all of any field's reserves in a single year.) If global oil demand were to grow by a modest 1% per year, though, the hypothetical supply shortage point moves up to about the year 2015. These calculations are not meant to suggest an actual prediction of a time when the world will "run out of oil." Even when the point is reached at which the reserves drop below 10 years of demand, oil production can remain very significant for many years. Rather, the estimates given here are meant to demonstrate that oil supplies can be expected to meet demand fully until at least about 2015, but that by the middle of the 21st century, it is likely that oil's market share will have to decrease.

KEROGEN

The acquisition of kerogen has proven to be a limiting factor for this resource. Not only does kerogen face all of the same environmental and production costs as coal, but it is also not usable in its extracted form and must undergo massive processing operations. Retort operations have failed commercially, and even at a price guarantee comparable to the highest prices reached in the

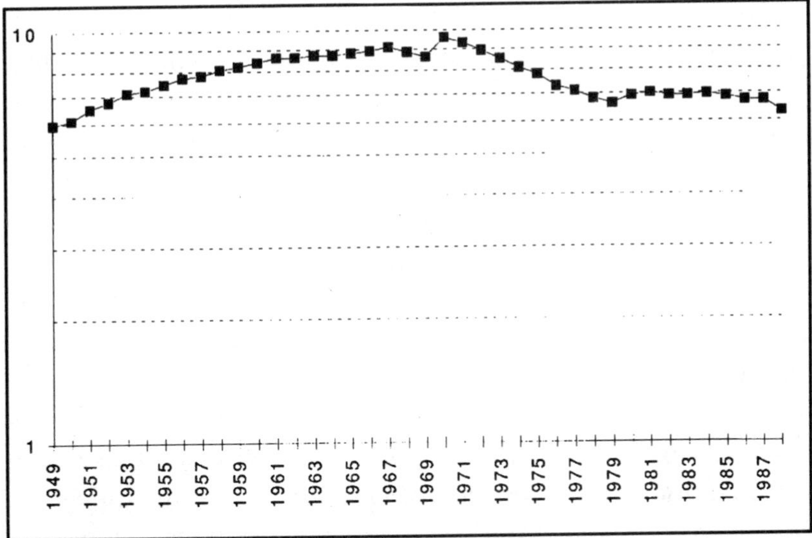

Figure 6–1 World petroleum reserves from 1949 to 1989.
Reserves only reached a peak in 1970. It would appear that a general trend of decling reserves is being established, at a rate of approximately 1% per year.
Courtesy Annual Energy Review, 1989, Energy Information Administration, DOE, Washington, D.C., Sec. XI, Table 14

late 1970s, this resource does not seem promising from a commercial standpoint. In addition, since the retort residue occupies more volume than the mined rock, kerogen is also undesirable from an environmental perspective.

BIOMASS

Acquisition of raw biomass, while technologically straightforward, is becoming an increasing human problem for the more than 1 billion people living in fuelwood-deficit regions. As women walk farther to collect their families' firewood, their hazards increase. Serious issues of injury in the cutting and gathering of firewood have been documented, as well as incidence of snakebite and rape by itinerant gangs.[6] It is, furthermore, the nature of firewood and charcoal acquisition that makes the direct use of biomass so environmentally damaging. The fact that demand tends to be location-specific and that transporting wood or even charcoal long distances is difficult, intensifies consumption in the immediate vicinity of population centers. One of the problems noted repeatedly in reference to the deforestation of England from the 1500s to the 1700s was that the demand was so tightly centered near cities, iron forges, and ship-building. These areas were quickly stripped of their trees, especially large-growth, older trees, while the consumers had virtually no incentive to extract wood from more distant forests. John Perlin's book, *A Forest Journey,* provides graphic details of the deforestation, including a quote from the famous author Daniel DeFoe that vast stands of trees could be found in some remote locations, which were "suffered to grow only because it was so far from any navigation that it was not worth cutting down." Lumber was used extensively for shipbuilding, but floating logs down rivers was the only effective way to transport them long distances for fuelwood uses as well.[7]

More sophisticated biomass utilization schemes invoke other problems. Some biomass resources, such as fecal material of range cattle, is broadly dispersed. While the anaerobic digestion of fecal material to produce methane can be done in an environmentally benign manner, even leaving a residue with soil-fertilizing potential comparable to the original material, the cost and effort of gathering it are prohibitive if the source is not already concentrated as in a dairy or feed lot. The modest contributions possible from generating methane from organic waste streams that are concentrated should probably not be overlooked. Indeed, it would seem that nonindustrialized countries seeking to build sewage plants to meet the press of urbanization can design those plants to optimize the energy production. This could provide some revenue benefits to developing sewage systems, in addition to the clear and present benefits to health.

SOLAR

Two problems with the acquisition of solar are largely responsible for limiting its use. The first problem is in the very nature of the resource: it is intermittent. The second problem is the low conversion efficiencies of existing technologies. The first limitation is unavoidable. It can really only be addressed through dramatic improvements in storage and transportation technologies, if solar is to become a dominant, stand-alone resource. There is reason for hope in these improvements, as they require developments of existing technologies currently being researched (such as the technologies of flywheels, of hydrogen electrolysis, or of super-conductors). While the technologies for photovoltaic and Solar Thermal Electric Conversion also exist, they are not in a state of development that holds large promise for engineering improvements. The developments that have brought prices down dramatically on solar energy systems are as much factors of economics as of technology. Prices have fallen as the manufacture of the devices has moved from prototype level to mass production. Once a unit is in mass production, cost gains cannot be expected to continue very long.

WIND

The application of wind power, like solar, is limited by intermittent occurrence and low collection/conversion efficiencies. The same potential for improved storage and/or electric transmission technologies offers promise for wind as well as for solar. Dramatic advances in conversion efficiencies, again, seem doubtful. One additional constraint faced by wind technologies is that high wind velocities are not attractive settings for population (and thus energy demand) centers. This last caveat, then, focuses on a greater need for better electric transmission efficiencies. One wind technology that has a history of successful application is in the transportation sector. It is limited to sailing ships rather than highway vehicles because the preponderance of automobiles would result in vehicles "stealing one another's wind" constantly. However, the importance of decreasing fuel consumption on the oceans makes sailing an interesting application.

HYDROPOWER

Hydropower has a long, successful history in terms of tapping the vast energy of flowing rivers. The large-scale applications require the construction of dams which flood huge areas upstream. They thus create serious ecologic disruption. The construction of many more large hydroelectric dams seems both unlikely and undesirable. The potential for this resource probably lies in the deployment of *micro-hydro* projects. Since a micro plant can sit on a river's edge, with its turbine gathering energy from the passing river's flow just as the earlier watermills did, it does not incur the ecologic cost of large-scale plants.

GEOTHERMAL

As long as the acquisition of geothermal energy is restricted to naturally porous, permeable steam or hot water reservoirs, this resource will remain a very small contributor to global demand. The potential for tapping the ubiquity of geothermal energy through *Hot Dry Rock Technologies* could expand the role of geothermal energy immensely. The technologies exist to produce electricity from geothermal energy in a closed loop, in which essentially all of the fluids drawn from the formation are recycled into it. Thus, geothermal energy can be produced with virtually no environmental impact. (The Hot Dry Rock Technologies would require a large water source to inject into the initially dry rock, but saline waters could be used and recycled in the closed loop.) The resource base is enormous and may be technically renewable for a very long time—perhaps as long as the life of the sun. To tap geothermal energy effectively in these terms, though, will require limiting production rates stringently, such that heat flux from the earth is comparable to the amount of heat withdrawn from the reservoir.

NUCLEAR POWER

The resource base of uranium 235 for conventional nuclear fission technologies is large. The potential is vast if *breeder* reactors are deployed, which use the much larger uranium 238 resource base, because neutrons emitted in the reactor core can convert the non-fissile uranium 238 into fissile products (mainly plutonium). Breeder reactors can produce less radioactive waste ultimately than conventional nonbreeder reactors as well as extend the potential resource base. Fears of nuclear accidents and concern about the proper handling of radioactive waste have brought nuclear reactor production to a screeching halt in the 1980s and 1990s. Ultimately, nuclear power must be compared in the near term to the other commercially significant resources and to resources with sufficient abundance to take on and maintain a large market share. Coal is the only resource likely to meet both of those requirements in the coming decades. Is coal better than nuclear power? Coal-generated power has essentially no risk of a single, catastrophic failure claiming lives of the general populace. However, the small risk of a nuclear accident should be balanced against the known hazards of coal-mining and the health effects of air pollution. There is no compelling evidence that nuclear power is worse than coal power.

Fusion is, at best, a resource for the medium- to long-term future. If the technology becomes viable, the resource base is nearly inexhaustible. The amount of energy released per atom in fusion is remarkably large. Nevertheless, the multimillion dollar laboratories around the world striving to develop a workable fusion technology have yet to achieve any sustained reaction that produces more energy than it consumes. (For an energy resource, this is a fatal flaw.) Fusion has the theoretical potential to provide energy on a fantastic scale, with much less radioactive waste than is produced with fission reactions. The fundamental question is whether the potential of fusion justifies the magnitude of research support it receives, relative to the development of other resources such as solar and wind.

TRANSPORTATION AND STORAGE

The ability of energy carriers to be transported to the consumer and to be available on demand are essential to the energy serving human needs effectively. This issue can be a matter of convenience as in being able to turn on a television whenever one wishes. It can also be an issue of survival, though, as in the case of providing heating to a dwelling in the winter. As energy-use progressed from firewood to coal to oil and gas, it also progressed towards maximizing the ease of carrying and holding energy. One person, carrying a five-gallon can of gasoline is carrying enough stored chemical energy to propel a 1,000 pound vehicle more than 100 miles, even considering the inefficiency of the use.

COMBUSTION FUELS

The combustion fuels all tap chemical energy stored in organic molecules. The purity of the organic content increases from firewood through coal to oil and finally natural gas, which explains the increasing energy density of these fuels. In addition to being the purest organic chemicals, oil and gas can be pumped through pipelines with great efficiency. This is one of several reasons that favor converting the solid fuels (biomass and coal) to a fluid prior to use. Coal slurries present an option as well as less efficient processes of liquefaction and gasification. The long carbon chains in biomass can be converted by microbial action to liquids and gasses; this can be accomplished either through fermentation to alcohol or anaerobic digestion to methane. Conversion to fluids is probably essential to effective and environmentally sound use of biomass.

ELECTRICITY

Essentially all of the noncombustion sources are converted to electricity for transportation. Electricity offers humanity a more versatile energy carrier than any other known and one that is almost perfectly clean at the point of use. Like piped fluids, electricity is available on demand. Like a pipeline, significant investments in infrastructure are required before a society can make use of the most convenient energy form.

An external cost of electricity that has come to attention recently is the potential cancer toll on people (especially children) living near the ultrahigh voltage power lines. The United States continues to study the possible connection between cancer rates and exposure to the ionized air of the corona surrounding power lines, while other industrialized countries have concluded that a hazard does exists and have begun studies to ameliorate it.

CONVERSION AND END USE

In this section, the role of consumer preferences in shaping demand are considered. The character of demand often dictates energy choices and consequently the set of costs incurred. Since the consumer commonly uses secondary or even tertiary energy carriers, the issues related to consumption do not break out effectively along the lines of the original source. The consumer generally does not care, or even know, whether his or her electricity is generated by a coal-fired, hydro, nuclear, or other power plant. Convenience is one of the highest consumer priorities. Coal was more convenient (because of its energy density) than firewood. Oil was more convenient than coal because of its portability. Electricity has become a highly favored carrier because of the convenience offered by its versatility. There are few tasks that electricity cannot perform (powering aircraft might be an exception). Electrocution does present a cost externality similar to the hazards of combustion fuels but one that has not noticeably deterred consumption.

One of the limitations on electricity that was ignored for some years when energy prices were very low is that there is an efficiency loss with each conversion from one energy form to another. The proliferation of homes heated by electricity is one problematic example. The cost of heating a home with electricity is definitely higher than heating with gas, oil, or coal, because the electric generating plant incurs the cost of the primary energy source (most commonly one of the combustion fuels just named) and generates the electricity at limited efficiency. The homeowners must then pay the price of the original fuel before their heating appliances convert the electricity back to heat, with the attendant efficiency loss.

Vaclav Smil in *Energy, Food, Environment* convincingly details the progress from one resource to the next. If he is right, it would suggest a shift in consumer priority to value environmental issues. This is not to say that consumers have never placed value on environmental factors. The history of coal development in England was a response to the loss of forests by firewood dependence. Those who first report experiments with coal-fired iron forges speak of their effort to replace firewood. Still, concern is focused on the utilitarian needs for the production of the forests.[8] England's impressive growth in energy consumption that attended the switch to coal was seen to imperil the local environment and human health. Hence, smokestacks and chimneys were built taller to allow the winds to carry the smoke beyond the local vicinity. One could dismiss this as being a purely self-interested step that bore no benefit to the environment as a whole, yet it was likely based on a notion that the global environment was so large that it could absorb whatever humans poured into it. Indeed, limited, localized views of environmental issues continue to be prevalent.

COMBUSTION

Coal, oil, gas, biomass, and kerogen-derived fuels all produce heat, water, carbon dioxide, and varying amounts of other emissions in releasing chemical energy. The overall efficiency of converting the chemical energy to the desired work is a function of the combustion process employed, the machinery, the number of intermediate steps in the conversion, the task, and the fuel itself. Automotive transport is a major consumer of combustion energy and an example of low efficiency. Since the internal combustion engines that power most vehicles use the mechanical energy of hot gasses expanding from an explosion, a large amount of energy is inevitably lost as heat in the process.

Stationary plants that employ combustion to convert water to steam for turning generators have the opportunity to employ cogeneration techniques, as do industries using heat for processing materials. Recovery of the waste heat has already provided great improvements in industrial energy efficiencies. All the commercial or natural combustion fuels produce carbon dioxide. Wood and coal produce the most per unit of energy. Oil produces a little less, and natural gas yields the least amount of the greenhouse gas. Furthermore, the emissions of particulates follow the same progression. Therefore, natural gas contributes less to atmospheric pollution than any other combustion fuels. Vaclav Smil's projections may be borne out, if informed consumers value the environment. Gas has frequently been bypassed in exploration because the demand for motor vehicle fuels placed a higher value on oil. Therefore, potential reserve additions of gas will probably exceed those of oil when the price discrepancy disappears, as it is even at the current time.

THE ROLE OF CONSERVATION

In the industrialized world (especially the United States), conservation must play a central role in the evolving energy mix. The citizens of the United States use twice as much energy per capita as do the people of other affluent, industrial economies such as Japan, Germany, and Sweden. People in these countries, though, use 40 times as much as the residents of the lowest income countries. Conservation is a consumer choice that can involve either relinquishing the benefits derived from energy use or investing in means to derive the same benefits from less energy.

Considering that transportation is one of the largest energy-consuming sectors in the United States and consumes energy at extraordinarily low net efficiencies, this sector demands special attention.

Voluntary Energy Thrift • A great deal of American energy consumption is discretionary. This includes unnecessary trips, solitary travel that can readily be shared with others, and travel that does not require a vehicle. There is tremendous opportunity for energy savings here. A large portion of American travel involves trips of less than two miles. Healthy, ambulatory people can certainly walk distances less than two miles. Indeed, the low impact exercise would doubtless be good for most Americans. While some short trips really do call for vehicular transport, because of the mass of materials being hauled, inclement weather, or urgency, eliminating even one-third of these automobile trips would mean a significant energy savings. This would require a change in attitudes and values but not necessarily any real loss of utility.

A less dramatic change in values, though, might lead to an increase in the use of mass transit, which can also reap large energy savings. Trips need not be forsaken if travelers choose to ride together. The lanes reserved for carpools in many urban areas are normally less crowded, and travel in them is faster than the other lanes. This provides positive incentive to move away from the single-occupant automobile, especially for those commuting to work.

Another value change that would save energy would be to place a priority on purchasing homes near work (or at least near mass transit). One of the reasons that the automobile consumes so much energy in the United States is urban sprawl. Many of the most sought after homes are in quasi-rural suburbs that may be quite distant from the workplace. If consumers chose their home based on its convenience to their work location, a great deal of energy could be saved. How can affluent workers be induced to return to the cities that they have abandoned to the laboring class?

Building Standards • Industrialized world building codes have called for increasing levels of thermal insulation which improves the efficiency of heating and cooling. Modern windows can have insulative properties comparable to those of the surrounding walls. In cool climates of the northern hemisphere, southern-facing windows are now capable of gaining solar heat during winter days without losing any additional heat during the nights. Therefore, window design has the potential to save considerable energy. It is possible, as well, to retrofit super-insulated windows in existing structures, sometimes replacing very inefficient windows.

In-line water heaters, used in buildings in Europe and Japan, can save thermal losses by maintaining the temperature of a 40- or 50-gallon hot water reservoir that is typical in America. They can also save thermal losses in moving the hot water from the remotely located tank to the point of use and save water as well in the process.

Appliances • In the industrialized world, consumer appliances such as refrigerators are major energy consumers which have considerable potential for savings. Improved insulation, lighter motor and compressor parts using synthetic materials, and *ramp-up* electronics provide much of the present savings potential. This is such a lucrative energy savings category that many states offer modest rebate programs to encourage consumers to purchase the most efficient new refrigerator models. One problem with the state programs can be seen by any consumer who inspects the EPA energy consumption stickers on several refrigerators of the same size. One may well find that the rebate is not available for a refrigerator that shows the lowest energy consumption tests, while it is available for less efficient models of the same or smaller size. This exemplifies one reason for many people's reluctance to trust government regulation policies.

Utility companies have also found that energy-savings rebates can be more cost-effective than investing in the capital outlay for large new power plants, and they create good publicity too. As well as rebates for replacing inefficient refrigerators, the utility companies often support rebates to consumers for replacing incandescent light bulbs with lower-wattage fluorescent bulbs. The modern, compact fluorescent bulbs can provide the same lighting as incandescents which consume more than six times as much electricity.

In the nonindustrialized world, though, the energy savings are not likely to be offered by new refrigerators or light bulbs. Cookstoves are a major target for improved energy efficiency in these countries. While it is true that improved, airtight woodstoves can operate four or five times more efficiently in the laboratory than open wood fires, and even field tests can show a doubling of efficiency, the substitution of stoves utilizing more efficient fuels should also be considered. A natural gas stove can achieve higher efficiencies in the field than an improved woodstove can achieve in the laboratory.

Savings from Decreased Raw Material Consumption • A great deal of energy is consumed in producing finished goods from raw materials, including metals from ores, paper from wood, bricks and ceramics from clays, and even synthetics from hydrocarbons. Efforts to increase the useful life spans of finished products and to recycle materials from products taken out of service have considerable potential. It is true that taking highly consumptive devices out of service to be replaced with new, more efficient devices can be more energy efficient than trying to sustain the longevity of the antiquated machinery. (The old trucks plying the roads in nonindustrialized countries would be a good example of devices to retire.) In the cases where equipment should be replaced, recycling the materials is the best choice. Additionally, though, designing products to be efficiently used for decades rather than years can offer substantial raw material savings, with attendant energy savings.

Conservation Policy Options • As opposed to voluntary measures, policy options here refer to those measures mandated by governments. These options include taxation and tax credits, as well as regulations of various kinds. In the final analysis, the difference between a tax and a regulation is more of form and of who gets the money than of the ultimate effects. If a limit is set on the amount of sulfur dioxide which an electric utility company can emit, or if a tax is imposed on utilities for each pound of sulfur dioxide emitted, the incentive to the company to reduce emissions is similar. The tax may be more flexible, because it provides a continuous incentive to decrease emissions, whereas the mandated limit only provides an incentive to reach that limit—further reductions have no value to the company. The regulatory limit is more absolute. If a producer fails to comply, it can be shut down. The tax has the added advantage (or disadvantage) of diverting money in the short term to the government, which can earmark it for needed research and development of new processes and construction of infrastructure (e.g., high-speed railroads, etc.), or it can find ways to waste the money. The public lack of trust in government would seem to be one of the factors favoring regulatory overtaxation policy.

Policy options to limit sulfur dioxide and carbon dioxide emissions of industry have been created, and there is discussion of strengthening them. Forests in most industrialized countries are at least hypothetically protected by land-use regulations and by the fact that significant portions of remaining forests are on public lands. Surface reclamation of strip mines has been mandated for years.

Most industrialized nations tax fuels much more heavily than does the United States. The taxation encourages conservation by increasing the effective price to the consumers. Policies can be directed toward corporate entities or toward citizen consumers, whether in the form of taxation or of regulation. Regulation of the public may be likely to raise questions of inhibiting personal liberty.

BALANCING THE IMPACTS

The use of any energy resource incurs a range of costs to society and the environment. Some of those costs are clearer than others—the health hazard to coal miners can be rather easily estimated, but the risk associated with long-term leakage of radiation from nuclear waste disposal sites is much fuzzier. Furthermore, how do human health hazards compare with environmental damage? Even with a totally anthropocentric view (one which values human welfare to the exclusion of any other life), environmental damage ultimately affects human welfare. The radio talk show star, provocateur who remarked "I've never seen a forest that wouldn't look better as a parking lot" might sing or croak a different tune if, indeed, the world's forests (or more generally growing vegetal matter) were depleted to the point that oxygen levels began to fall.

Energy consumption brings very real benefits to human life. On a very basic level, it permits cooking food, heating living space, and lighting the dark. These uses are largely nondiscretionary. Since human existence depends on such basic energy uses, and since survival is a compelling instinct, there is actually a negative environmental impact to providing people with less energy than is required to meet their basic needs. The survival instinct will drive people to destroy anything around them in order to meet those needs. People will cut down forests, strip bark off of trees they cannot cut down, and gather dung to burn which should go to fertilize their fields (the equivalent of eating one's seed corn). It clearly imperils next year's survival, but the need to provide for today's survival is incontrovertible.

At a less fundamental level, energy still provides very real benefits in terms of transportation, preservation of foods and medicines, communications, enhancement of building materials, and even entertainment. It is possible to conserve energy by curtailing use in these discretionary tasks, but there is a loss of utility. It is also possible to limit the environmental impacts of energy consumption by utilizing new technologies for more efficient energy conversions or by deploying means to remediate the environmentally damaging aspects of energy use. There are monetary costs involved in developing and producing new technologies. As mentioned above, though, the money spent on improving technologies has a tendency to create economic growth.

Each step in the production and utilization of energy from any source carries with it a variety of impacts on human health and safety, the atmosphere, the biosphere, waters, and aesthetics. Policy analysts are wont to characterize the impacts as costs. This permits some assessment of the overall impacts across the broad range of energy sources, each with greater effects in different sectors. In this approach, one classifies all of the negative impacts as costs and the positive as benefits. Direct costs are the easiest to assess. The fossil fuels provide the cheapest energy in general at the present time. The direct costs of the devices to convert solar or wind energy to a useful form are very high; although the energy supply itself is free, it would take years for fuel savings to equal the cost of the device. Another way to look at this is that the cost of fuel is so low that it is negligible in comparison to the capital cost of an energy conversion device such as photovoltaic cells, STEC, or wind turbines.

THE VALUE OF EXTERNALITIES

It is not feasible to achieve environmental benefit without assigning a value to it and paying that value. Then how much is a reduction of acid deposition worth? How highly valued is a reduction in carbon dioxide emissions, or deforestation? How much is it worth not to produce radioactive wastes, or toxic chemical wastes? Effectively, these are the questions consumers/voters must answer in determining the value that they want to pay, as the increased prices resulting from regulation or taxation are passed through to them. It would seem from historical consumer trends that the average American is not inclined to pay an energy premium to reduce environmental impact.

It is true, however, that energy prices, as seen by the consumers, have not continuously risen in response to increased regulations (in constant dollars). In some cases, the regulations can promote increased efficiency which eventually pays for itself. In other cases, new benefits are provided, such as increased financial security of workers and circulation into the general economy of the gains received (such as in the case of compensating miners for risk and injury). In fact, the economy, like the global environment, is a complex, dynamic system with numerous feedback loops.

In addition to fostering a societal commitment to preserving the environment and to internalizing externalities, it is important to assign costs equitably across the board. This is to say that since each energy resource has a different mix of externalities, it is possible to create a harmful market imbalance between resources. If a value is assigned to reducing acid rain (perhaps through regulations limiting the amount of sulfur permitted in plants, or through a "sulfur tax"), regulations are passed favoring nonsulfur-bearing fuels. Might an incentive thus be created to convert to wood fuel? This has happened, and eventually, it could create the negative outcome of causing forests to be cut to provide low-sulfur fuel to replace coal. (It has been demonstrated previously that intensive fuelwood plantations are ecologically nothing like true forests, and that the total emissions from direct biomass combustion are problematic.) Therefore, paradoxically, the regulation to reduce acid rain in order to preserve forests and lakes could have the effect of destroying forests.

Of course, this is an unlikely example in most industrialized countries because there are regulations controlling the cutting of forests. That is precisely the point, though. To ensure that a regulation (or tax) has the desired positive effect, it must be balanced by comparable regulations on other negative impacts that other resources produce.

Finally, it is important that consumers (and policy-makers) have adequate information about the range of impacts of all energy options and the severity of those impacts. The information is significant not only in assigning relative values, but it is also critically significant to recognize the costs and potential costs of resources that are not well tested. While the grass may always seem greener in a new field, it may not stand up to the tests of intensive consumption. Negative impacts are rarely easy to foresee.

It can be argued that the reason fuel prices are as low as they are is that many actual costs remain external to the energy-producing companies. If these externalities were included, the price of energy would rise. In essence, the argument has been seen to be true through past experience. When coal companies did not have to provide for miner safety and were not compelled to pay for deaths and disabilities in their mines, these costs were kept external. The mining companies did not have costs for safety and insurance payments to pass along to the consumers. (And all costs are passed along to consumers.) When reclamation was not required, the costs of damage to the land's surface remained external. When oil companies did not have to pay damages incurred from oil spills, those costs remained external.

As new regulations required companies to pay each of those costs, they became internalized and thus a part of the price paid by the consumer. Today, it would be inconceivable to operate a mine in which the miners were not compensated for their risks and actual injuries. Every new set of regulations seeks to internalize another externality. Of course, a large number of people in industry would reply that the energy industry has been very heavily regulated, internalizing many costs, and that every cost internalized is passed along to the consumer. How many Americans will say that they want to pay more for oil? It is certainly true that new expenses placed on the companies will ultimately be reflected in higher prices. Price controls can forestall that reflection until a supply shortage drives the priority to find and produce more of the resource, which demands higher prices to permit larger, more exotic exploration and production efforts. The basic law of supply and demand ultimately works. The 1970s showed what seems to be the logical result of price controls: a skyrocketing price that makes up for all the expenses not passed along earlier when shortage causes the controls to be removed.

There is no way to get around the reality that if consumers want to place a value on environmental preservation and/or on developing competitive alternative energy sources, they (we) must be willing to pay the value they place on it. It may be argued that some alternative energy sources are very close to being commercially competitive. If these resources do not have high environmental costs, then including environmental externalities in the energy price may allow the new resources to displace existing ones without a large price increase. If photovoltaic cells, for instance, were just barely below the competitive price of combustion energy sources, a slight price increase might bring them solidly into the market place. If coal-fired electric utility plants had to pay the price of restoring the pH balance of lakes and lands within a 100 mile radius, that would probably be a very large cost factor to pass along to the consumers. However, if photovoltaic cells can be cost-effective at a very slightly higher price, then photovoltaic cell production and sales would soar, finally replacing the need for that power plant. If the utility company were given a timetable to effect cleanup, so that costs were incurred gradually, the transition might even be made smoothly, and give the burgeoning photovoltaic industry time to build adequate capacity and to take up jobs lost in the failing utility industry.

What will happen if the price and environmental impact assumptions about an alternative energy source are incorrect? If any factors cause the alternative resource to fail to compete at a small price increase, then energy prices will continue to rise, even to the point at which all of the cost placed on the utility is passed along to the consumers. If experts estimate that a given technology could compete at a certain price, why could they be wrong? The clearest potential for error comes if experts extrapolate prices based on assumptions about technological advances that are not yet in hand. Those advances may be years in the future or may be impossible. Another source of error is that a required raw material may be in limited supply and become costlier as demand for that material increases. The new technology may involve a hazard that has not yet been identified or adequately taken into account. (That was certainly the case for nuclear power.) So, a wise consumer would advocate regulations that internalize externalities to the price level at which the consumer actually values that externality. The alternative energy on which a consumer counts may fail to compete even with new regulations. The consumer may end up paying the full cost that a new regulation assigns to a previous externality.

THE ENVIRONMENT AND RENEWABILITY

The terms, *environmentally sound* and *renewable* are often used interchangeably in reference to energy as if they are synonymous. However, is there any real connection between the rate of a resource's renewal and its environmental impact? It would seem to follow that if drilling in Iceland someday proves that oil and gas are not fossil fuels, but are actually renewed through metamorphic rock processes at global rates exceeding consumption, then oil and gas will become more environmentally sound. (Wells are drilled in Iceland where classic theory says that no oil could occur. Finding oil there would confirm the opposing hypothesis that oil and gas are formed by rock metamorphosis, which could happen on a larger, more renewable scale than the decay and alteration of dead things.) Of course, the environmental impacts associated with oil production and consumption are completely unrelated to its rate of renewal. In reality, renewability is an economic issue. How long will a resource last at a given consumption rate, if it is not significantly renewable?

A financial analogy will illustrate. A billionaire is an extraordinarily wealthy individual. If the billion dollars is invested at a minimal 2% interest, the investor has a renewable income of 20 million dollars per year. That rate of renewal would provide a permanently sustainable resource for nearly anyone. (Similarly, if another individual with only one-tenth the wealth were able to find investments paying 20% interest, the annual income would be the same. The second investor has a higher renewal rate.) The first question to ask is: Does a high renewal rate guarantee that an individual never goes broke? History has shown countless examples of people with incredible wealth who are able to squander it. The rate of consumption of a resource (whether that resource is money or oil or firewood) is as important as the rate of renewal.

Biomass resources closely fit the analogy. Like money, the amount of renewal is a function of the size of the account. If half of a monetary resource is squandered, the amount being renewed decreases by a similar factor—so too does the renewal of biomass as forests are lost. As it would be senseless to tell a billionaire who has spent half of his fortune that the income from his fortune is more than anyone could spend (since he has already spent far more than that in order to lose half of his wealth), it is senseless to speak of the renewability of biomass in a world that has lost half its forests. In the case of a billionaire who has lost half his billion dollars through overconsumption, it remains easily conceivable that improved management of his consumption would permit him to live very comfortably on the earnings from his remaining fortune. The same may not be true for woody biomass on a global basis. The wealthy earth has 5 billion dependents drawing on her wealth. While that wealth was once immense and still remains vast, the amount of nondiscretionary consumption alone may already exceed the income generated from the remaining wealth. If that is so, then limiting consumption alone cannot balance the budget.

The earth's wealth, though, is diversified. Consumption of biomass can be diverted to other energy sources. Since the biomass resource base is critical to the earth's environmental integrity, it is probably wisest to continue displacing the use of wood for energy with other sources, even fossil fuels, to permit the annual income generated by the remaining forest reserves to rebuild the depleted wealth.

The renewal rate for fossil fuels is much less than the renewal rate of biomass because only a small portion of plant growth is annually deposited in the long-term account of fossil fuels. Certainly, then, fossil fuel use depletes a reserve with a much more limited renewal rate than exists for biomass. Fossil fuels will someday be depleted and unavailable for all practical purposes. However, the consumption of fossil fuels does not deplete the account (reserves) from which renewed resources are generated. Traditional biomass consumption does.

The Closed Carbon Cycle • It is undeniably true that life is supported on the earth by a carbon cycle, in which growing plants take up atmospheric carbon dioxide and, with the energy input of sunlight, convert the carbon dioxide into free oxygen and complex carbon molecules. The other side of the cycle is the oxidation of those carbon molecules by living organisms, producing carbon dioxide, which is released in respiration. Additionally, of course, some of the carbon molecules originally formed in photosynthesis are broken down to carbon dioxide through rapid oxidation (burning). If a point is reached at which the balance of the cycle is disrupted, and there is not as much oxygen being returned to the atmosphere by photosynthesis as is being taken up by oxidation processes, the atmosphere will be depleted of oxygen and the results will clearly be catastrophic. Since a large tree produces about 2 kg of oxygen during daylight and humans consume about .8 kg per day, if global vegetation does not exceed the equivalent of 200 million trees, there will not be enough oxygen even to support human breathing. But the equivalent of another 2.2 billion trees is required to replace the oxygen consumed by combustion fuels.[9]

This extreme scenario is much clearer than the greenhouse effect. There can be no question of how serious the effects will be or what they will be; animal life (including humans) will be imperiled and die of asphyxiation until a global balance can be regained. Is this outcome likely? No, but it is presented to make clear that there is a point at which a limit to human and industrial growth at the expense of the environment unquestionably imperils human survival. Indeed, it is this author's opinion that the catastrophic scenario described will never occur because less obvious negative feedback mechanisms will be or are being activated long before growth reaches the point of global oxygen depletion.

ECONOMIC CONSTRAINTS

Generally, consumers of energy have responded quickly and strongly to changes in energy prices. A good deal of that response has been in the form of protest and accusations against the rising prices and companies benefiting from those increases. A more direct market response has also been readily apparent in consumption patterns.

The Cost of Fuel Efficiency Standards • Mandated vehicular fuel efficiencies drew added attention in the United States during political debates in the 1992 election campaign. One side of the debate advocated phasing in increasing fleet fuel efficiency standards over the next several years. The other side of the debate maintained that requiring such improvements would cost the American economy billions of dollars. But in fact, persisting in the trend towards inefficient transportation cost America approximately 8 billion dollars in 1992 alone, compared to achieving a fleet average fuel efficiency of 35 miles per gallon. (Since this would be a savings of imported crude oil, it would be a savings of foreign exchange. Foreign exchange savings are real savings to a nation's economy, while expenditures within the economy can contribute to the nation's economic growth.)

Similarly, some serious questions can be raised about the figures assigning a large cost to the transition to higher automobile efficiency standards. The materials for smaller, lighter vehicles cost the manufacturers less to buy, and no significant new technologies are required in order to meet fuel efficiency goals. Most manufacturers already make some models that exceed projected standards. Their cost would simply involve modifying the relative production of existing models. Since auto manufacturers make annual model changes, this would seem to be a minimal cost to them. Would policies to improve average fuel economies include mechanisms to prevent American consumers from simply buying an increased share of inefficient foreign cars with the horsepower and options desired? If so, then more stringent standards might benefit a nation's economy.

THE ENVIRONMENT AND AESTHETICS

Substituting subjective issues of aesthetics for substantive issues of the environment has not only drawn attention away from solid policy analyses but has also opened the door to debates which trivialize environmental issues. It is this substitution that leads to rhetorical arguments against environmental policy, such as "How much is it worth for me to be able to see across the Grand Canyon?" It is obvious at even cursory analysis that what one sees or doesn't see is an issue of aesthetics with very questionable value beyond the individual. However, in this example, if one becomes unable to see across the Grand Canyon, it may well be a symptom of the severity of an environmental problem—air pollution.

Offshore drilling is a more common example of confusing these issues. When oil-producing companies attempt to bid for offshore leases, coastal residents often raise objections couched in environmental terms, mixing issues of environmental damage with issues of personal aesthetic judgments. On the one hand, discussion of the fragility of a local ecosystem is truly an environmental issue. If extraordinarily dense populations of marine life or endangered species exist in the vicinity of a proposed offshore drilling project, a legitimate imperative can be issued to study the likely impact of normal operations on the marine life and precautions set forth to respond to blowouts to protect the rich estuaries, which are likely to be on the shoreline. On the other hand, an individual with a home on the coast who does not wish to look out his or her window at an offshore platform is not expressing concern with impact to the environment. Indeed, as long as that individual consumes commercial energy at or above the level of the average American, he or she is demanding oil that is produced off of the continental United States. If that demand is met by imported oil, the likelihood of catastrophic spills of oil increases dramatically over the likelihood introduced by offshore, domestic operations. So, it would seem that if the coastal resident is concerned about the environment, curtailing his or her consumption levels would be the logical solution or advocating domestic offshore production.

Granted, the real issue may not be either environmental or strictly individual aesthetics. The individual may have purchased a very expensive home on the coast, with a pristine view of the ocean's natural majesty. Perhaps the potential future buyers of the property would not be inclined to pay as high for a view of an offshore platform. Whether an individual's property on the coast is devalued by some disruption of the ocean view, though, is not a real environmental issue. The individual is free to object on the grounds of decreased property value and may gather enough concern from neighbors to persuade the company to employ subsea completions and pipeline, which would leave the vista undisturbed once drilling is complete. Couching the argument for maintaining personal wealth in environmental terms, however, seems to distract attention from, if not diminish, real environmental issues.

ECONOMIES OF SCALE

The expression *economy of scale* normally refers to unit cost savings available to large-scale enterprises. Essentially all economic activities have certain fixed costs that must be borne regardless of the size of the project (e.g., salaries and travel expenses for managers and specialists going to negotiate with the original owners of the raw materials and to evaluate the project site). There are also many expenses that are not proportional to the size of the undertaking (e.g., moving equipment and personnel to a remote drilling site). Consequently, these fixed costs can be amortized (spread out) across more units of production for a large project than for a small one. However, a large project requires a larger capital investment and concentrates the environmental and social impacts.

The issue of capital requirements is particularly significant to lower income countries. Capital is scarce. Demand levels and the ability to pay are modest. The ability to distribute large start-up costs across an even larger demand base is generally absent. This fact deters foreign investment for developing resources in lower income countries. For instance, the limitations of local markets dictate that petroleum produced in most of these regions is destined for export to the large markets in the industrialized world. This also means, though, that only very large reserves are developed, so that there is enough production to absorb the immense costs of establishing tanker-loading facilities. The size required for an oil or gas discovery in many lower income countries to be commercial under these constraints is about 100 times larger than would be required for a commercial discovery in the United States. A great deal of oil and gas is left untouched, then, in countries that may be paying half of their foreign exchange for energy imports.

Small-scale, localized production might enable many lower income countries to produce their indigenous oil and gas resources to replace diminishing firewood resources and to reduce the foreign trade deficit. The small-scale oilfield technologies are not often presented to the people seeking energy options in lower income countries. In 1986, the United Nations Institute for Technology and Research hosted a conference to explore the possibility of attracting (independent) American oil companies to bring their small-scale, localized operations to countries in need of energy. The findings of the conference included the recognition that whether large or small, companies need to make a profit, and American companies want to be certain of bringing their profits home (soft and weak currencies can make this a real problem). There was also concern about finding adequate markets to optimize production of the oil and gas discovered since lack of infrastructure can make it difficult to reach markets. Companies also wanted assurances that their operations would not be nationalized. These are important issues for any profit-seeking corporation.

Another alternative was founded in 1987 at the University of Rochester in New York (by the author and colleagues). That project, called Access to Hydrocarbon Energy for African Development (AHEAD), found that a public charity could develop resources cost effectively without any of the constraints for profit-seeking corporations noted in the UNITAR conference. AHEAD has found that officials in countries such as Ghana and Mozambique are well aware of the potential benefits of developing their indigenous natural gas resources and have taken several innovative steps toward that goal. The idea of AHEAD is that energy, like food and shelter, is essential to human survival and that establishing natural gas production and building infrastructure are critical for lower income countries. It may be most appropriate for a nonprofit organization to facilitate a first step.

Other small-scale activities that may be significant to people in nonindustrialized countries include mini-hydroelectric projects, as well as several variations on solar and wind technologies. Small-scale hydro projects seem to have potential because their capital costs are manageable, while they can provide versatile electricity with little environmental disruption.

A number of solar projects have failed because they were not adequately matched to the consumer needs, such as the parabolic cooker project discussed earlier. That solar project failed to consider that all adults in the region worked from dawn 'til dark in the fields and did all of their cooking after dark. Solar box cookers require very little capital investment and are conducive to unattended crock-pot-type cooking, but many of those available do not work well. A Ghanaian official explained to the author that well-intentioned projects will always garner positive reviews from the local clientele because politeness prevents people from complaining about a device they are given, no matter how worthless it is. If engineers get together with social scientists on the design of such devices to meet both the dictates of consumption patterns and of physics, solar cooking devices may be a significant contributor to one of the most basic energy needs.

CONCLUSION

Planning for the future requires defining time horizons. The appropriate plan for tomorrow may be very different from the appropriate plan for a century hence. The further one tries to plan into the future, the less certain are one's assumptions. Plans extending a millennium into the future would probably not be worth cutting down the tree for paper on which to write the plans. It would be implausible to predict the technologic developments which would alter the constraints dramatically. Will safe, controlled fusion be developed? Will a breakthrough be made in photovoltaic cell design? The history of the development of humanity suggests that both of these developments are very plausible, IF human needs demand them. However, the path selected through the coming years and decades may well impact which opportunities for development are available in the distant future.

For the purposes of this discussion, time horizons reflecting the near term refer to immediate actions extending to approximately a decade hence. The medium term is considered as the current generation's and their children's lifespans. The long term is considered to be three or more generations hence.

Options that are viable contributors to the near-term energy mix in the industrialized world include 1) conservation, 2) coal, 3) oil, 4) gas, 5) small-scale solar, 6) small-scale hydro, and 7) small-scale wind. The options are essentially the same in the nonindustrialized world except that net conservation is probably not feasible in the poorest countries that enjoy very few benefits of energy consumption, and biomass is a large part of the existing energy mix. Transfers of technology will be required for the nonindustrialized world to increase significantly the share of any of the nonbiomass resources.

Conservation is likely to be an important part of the energy mix for the industrialized world, certainly in the medium term. Conservation options for the medium term may include population control measures. Clearly, the fewer people being supplied by the earth's resources, the less demand is being placed on those resources. Although population growth rates in lower income countries are much higher than industrialized countries, population growth control, if adopted, should be considered for both types of countries. The rationale is that, although industrialized countries have a slower population growth rate, the energy-consumption rate is much higher. Conservation for nonindustrialized countries should probably include shifts to more modern, efficient energy resources and perhaps advance the planning to control growth in consumption.

Fossil fuels will undoubtedly continue to play an important role in the industrialized world's energy mix and the potential of nations to develop industrially and/or economically. Natural gas has been the least utilized of the combustion fuels, which means it is left with a large potential. Because of its clean burning nature and efficiency, its use may grow in the near term. The depletion of oil reserves may begin to constrain supplies as time approaches the medium-term marker. Until oil supplies are constrained, it would seem likely that the energy density of the liquid fuel will cause it to maintain a central position in the industrialized world's energy market. Coal's abundance will likely cause it to continue to maintain a major portion of its existing market share and receive continued attention as the source of synthetic fluid fuels. Kerogen's vast resource base probably will not be sufficiently important to revive interest in this costly energy form.

Geothermal energy has the capacity to expand its market substantially. It can be produced with minimal environmental impact, but some of the most prospective regions for implementing the prevailing geothermal technologies are scenic areas in restricted parks. It does not appear likely to gain much market share before the medium term.

Hydropower may have reached the zenith of its large-scale applications. Small-scale applications are being discussed but will probably not be widely employed in the near term.

Solar and wind energy both receive considerable press and did make gains during the 1970s and 1980s. Large-scale applications are still not cost-competitive with the fossil fuels or hydroelectricity unless they are supported by policy actions. Taxes or regulations imposing a larger share of environmental costs on the established energy producers might enable solar and wind resources to gain a market preference. The transition to providing a large share of the world's energy, though, will require large industrial output of the necessary equipment. This will require building new factories, as the equipment is not currently in mass production. Furthermore, as the new technologies become more extensively distributed, costs yet unseen may emerge.

Effective nuclear power moratoria in some countries, most notably the United States, will almost certainly limit the growth of nuclear energy production into the medium-term time frame. Perhaps by the time significant new nuclear plant construction resumes in the United States, breeder reactor technologies will be ready to take over. Fusion technology will almost certainly not be ready to take a large market share until well into the 21st century—if then.

Projections of energy mixes into the medium and long terms become increasingly speculative. By some point in the medium term, though, it seems clear that oil and gas reserve depletion will demand that other resources be available to replace their dwindling production rates. Coal certainly has the capacity to meet that need in the medium term, and the technologies to convert coal into versatile, efficient, and relatively clean liquid and gaseous fuels exist. Doubtless, coal will serve an important role as a *backstop* resource. The technologies for efficient exploitation of Hot Dry Rock geothermal sources may have evolved. In this author's opinion, the renewable energy source most likely to become a major producer is solar, because areas with a great deal of sunshine are attractive population centers while areas of high wind are not. Additionally, the electrochemical technologies of photovoltaics seem more likely to benefit from technical breakthroughs than do the mechanical technologies for harnessing wind power. However, breakthroughs in electric transmission technologies (such as superconductors) may obviate the need for on-site or near-site generation of electricity.

Almost certainly, within a couple of centuries, the resources used for combustion today will serve primarily as feedstocks for synthetic products and no longer as combustion sources. The transitions to provide for humanity's energy needs will no doubt continue to evolve. They are likely to seem as dissimilar to prevailing technologies as an ox turning a grist mill seems dissimilar to the computer on which this text is being composed. One thing is certain: energy will remain central to human life and development.

Cost and Benefit Comparisons

ENERGY SOURCE	COSTS	BENEFITS
Coal		
Environmental	Air pollution Surface mining damage	Pollution abatement techniques possible Reclamation is possible
Health	Underground mining hazards	
Direct cost		Low monetary cost
Suitability	Limited convenience	Potential for liquefaction/ gasification
Future		Abundant
Oil/Gas		
Environment	Hazard of oil spills in transport Carbon dioxide emissions Alleged hazards of offshore drilling, but limited evidence of serious damage	Well established and proven Cleanest of combustion fuels
Health		
Direct cost		Low monetary cost
Suitability		Versatile Transportable
Future	Depleting	
Biomass		
Environment	Deforestation Air pollution	
Health	Collection hazards Smoke in cooking areas	
Direct cost	High in deforested regions	
Suitability		Established
Future	Depleting in many regions	Potential for biogas from waste
Geothermal		
Environment	Risk of toxic salt release	Potentially very clean
Health		
Direct cost		
Suitability	Limited access Efficiency constraints	Good for direct heat Steam readily generates electricity
Future		Large potential resource base
Hydro		
Environment	Large-scale dam disrupts ecology	No air pollution
Health		No adverse effects
Direct cost	High capital for large-scale	Small-scale applications possible No fuel costs

Suitability		Efficient electric generation Efficient for mechanical work
Future	Environmental and location constraints	Not depletable
Solar		
Environment	Potential for toxic waste disposal (PV and STEC)	No air pollution
Health		No adverse health effects
Direct cost	High capital cost (PV and STEC) Large capital for construction (STEC)	No fuel costs
Suitability	Many applications not proven Requires storage or backup system	Passive space heating is proven
Future		Not depletable Vast resource base
Wind		
Environment	Hazard to birds	No air pollution
Health	Noise pollution may be an issue	
Direct cost	High capital costs for electric generation	Lower cost for mechanical work
Suitability	Not established Requires storage or backup system High wind speeds required (electric)	
Future		Not depletable Vast resource base
Nuclear (fission)		
Environment	Hazardous waste disposal	No normal air pollution
Health	Risk of catastrophic failure	
Direct cost	High capital cost	
Suitability	Requires high technological infrastructure	
Future		Abundant (especially for breeder reactors)
Nuclear (fusion)		
Environment		Probably nonpolluting
Health	Risk of catastrophic failure	But less radiation than in a fission plant
Direct cost	High capital High research	
Suitability	**Not proven feasible**	
Future		Abundant

*PV=photovoltaic; STEC=Solar Thermal Electric Conversion plants

A final energy-use decision should not be made by counting listings of costs vs. benefits for each energy source. Some cost categories are very high while others are relatively low. Some costs are important for some applications and insignificant for others. An effort beyond the scope of this book is required to assign value levels to each cost and benefit. When that is done, a much expanded version of the preceding table will be required to separate different types of air pollution for instance, and to separate different subsets of each type of energy, and to establish priorities for various applications. It should be made clear that the complex set of costs and benefits and the range of energy options require an analytical approach that is broadly comprehensive. However, the preceding table should offer some starting point for a practical means to weigh the merits and problems associated with each energy category and begin to plan an evolving energy mix for all of earth's societies.

ENDNOTES

1. Nef, J.U. 1932, *The Rise of the British Coal Industry*, George Routledge and Sons, Ltd, London, pp. 19, 20.
2. Smil, Valcav and Knowland, William E. 1980, *Energy in the Developing World*, Oxford University Press, Oxford, p. 91.
3. ibid.
4. National Research Council 1984, *Diffusion of Biomass Energy Technologies in Developing Countries*, National Academy Press, Washington, D.C., pp. 37, 38.
5. Presentation by William Simon, USGS, at the "Energy: A Key factor in Development and the Environment" conference at the University of Rochester, November 1990.
6. Baah-Boakye, Esther 1994, "Keeping Ghana Green and Energy-Efficient", *EcoJustice Quarterly*, Winter 1994.
7. Perlin, John 1991, *A Forest Journey*, Harvard University Press, Cambridge, MA, p. 221.
8. Perlin, *A Forest Journey*.
9. Ramage, Ann 1983, *Energy: A Guidebook*, Oxford University Press, NY, p. 106. Smil, Vaclav 1987, *Energy, Food, Environment*, Clarendon Press, Oxford, p. 251.

Appendix

Coal Ranks and Properties

	Moisture Content	Volatile Matter	Heating Value (Btu/lb)	Appearance
Lignite B	high	high	< 6,300	brown
A	high	high	6,300 - 8,300	brown
Subbitminous C	moderate	high	8,300-9,500	dull black
B	moderate	high	9,500-10,500	" "
A	"	"	10,500-11,500	" "
Bituminous, High Volatile C	moderate	high	11,500-13,000	black
High Volatile B	moderate	high	13,000-14,000	"
High Volatile A	"	> 31%	> 14,000	"
Medium Volatile	low	22-31%	69-78% carbon	"
Low Volatile	low	14-22%	78-86% carbon	"
Anthracite (semi)	low	8-14%	86-92% carbon	shiny black
Anthracite	negligible	2-8%	92-98% carbon	" "
Anthracite (meta)	"	< 2%	> 98% carbon	very shiny black

The coal types are listed in order from lowest to highest rank. The numerical values are used in classifying coals; the verbal descriptions are provided as general information. The changes described qualitatively are gradational as well as those for which numerical values have been assigned to represent the gradation. Note that from medium volatile bituminous through meta anthracite, carbon content is given, rather than heating value per pound, because the heating values level off above the higher bituminous ranks, and even decreases slightly in the anthracite ranks to an average of less than 14,000 Btu/lb for anthracite. Anthracite is favored for its exceptionally clean burn.

The classification scheme shown is the American system; the International System divides coals into hard and soft groups, in which the hard coals have heating values of greater than 10,260 Btu/lb, and the soft coals have lower heating values. The division equates to the lower range of high volatile bituminous C coal in the American system. Hard coals are assigned three-digit numbers, in which the first number reflects the class of coal, determined by the amount of volatile matter, up to 33%, and the heating value for coals with more than 33% volatile matter. The second and third digits relate to caking and coking properties respectively: the coal's behavior under different heating conditions. Soft coals are assigned four-digit numbers, with the first two digits reflecting the moisture content, and the last two digits reflecting the amount of tar in the coal.

Source: American Society for Testing and Materials, publication D388, p. 223. Moore, E.S. *Coal*, pp. 93–135.

FLUID FLOW IN POROUS MEDIA

A series of equations describing the flow of fluids through porous media was developed by Darcy, and they are called the Darcy equations:

$$Q = \frac{kA\ \Delta p}{\mu \ln (r_e/r_w)}$$

This is the most commonly used, radial flow equation. It is appropriate for the case where the areal extent of the reservoir is much larger than the thickness, the flow is laminar (non-turbulent), and the fluids are relatively incompressible (oil and water).

- Q = the flow rate—commonly in barrels per day for oil
- k = the permeability—in darcys or millidarcys
- h = the thickness of the producing interval
- p = the pressure difference between the average reservoir pressure and the wellbore pressure—normally in pounds per square inch (psi)
- μ = the viscosity of the produced fluid—in centipoise
- r_e = the radius of the area being effectively drained by the well, a distance at which the average reservoir

pressure used in Δp calculations should be encountered, commonly taken as the radius of the well spacing—commonly in feet

r_w = the radius of the wellbore, commonly in feet

If the above units are used, including millidarcies, the right side of the equation would be multiplied by the conversion factor of .00127

In the case of gas production, the Δp term is replaced by the difference of the pressures squared: $P_r^2 - P_w^2$. Additionally, the compressibility must be accounted for, with a formation volume factor (B_g). A comparable formation volume factor for oil actually does not particularly account for the minimal compressibility of oil, but for the shrinkage of oil as gas comes out of solution upon production.

THE ARCHIE EQUATION

This equation relates porosity and water saturation to the electrical conductivity (or its inverse, resesitivity). In its most commonly used form, it is arranged so that one solves for water saturation from resistivity and porosity measurements:

$$S_{wn} = \varnothing^m R_w / R_t$$

S_w = the saturation of water as a fraction of the volume of the formation

n = an exponent relating to the continuity of the water phase

$\varnothing$ = the porosity of the rock

R_w = the electrical resistivity of the formation water

R_t = the measured (true) resistivity of the formation

PIPELINE FLOW

For gas

$$Q_b = .0775 F_{tf} F_{pv} F_{gr} F_d F_t F_f ((P_1^2 - P_2^2)/L)^{.5}$$

- Q_b = the flow rate in thousands of cubic feet per day
- F_{tf} = the temperature correction factor
- F_{pv} = the supercompressibility factor (for non-ideal mixtures of gas at high pressure)
- F_{gr} = the gas gravity (density) correction factor
- F_d = the line diameter factor
- F_t = the transmission factor, which accounts for partially turbulent flow
- F_f = the drag or friction factor
- P_1 = the upstream pressure (in pounds per square inch)
- P_2 = the downstream pressure
- L = the length of the pipeline (in miles)

Other corrections may be required to account for measurements made at different temperatures and pressures, as well as for partially turbulent flow.

Source: *Steady State Flow Computation Manual for Natural Gas Transmission Line*, published by the American Gas Association, 1964.

HYDRODYNAMIC HEAD

The head, or height of a water column, determines the force it exerts, and thus the potential energy available to be harvested per gallon of water. The flow rate of the water multiplied by the pressure gives the energy output per unit time.

In the case of fresh water, in English units, the horsepower produced by a hydropower system can expressed as:
$Hp = .00189\ Q\ h$

Hp = the horsepower
Q = the flow rate (in cubic feet per minute)
h = the water "head" (in feet)

Since 1 Hp = 746 watts, the equation can be rewritten as follows for electrical output:

$$W = 1.4\ Q\ h$$

Efficiencies factors must be included to determine actual energy available to do work.

INDEX

Acid rain, 237-242
Automotive fuel economy, 159, 160, 183, 184. 210, 211

Biomass
firewood and charcoal, 28, 79-82, 123, 124, 161-165, 215, 239, 268
municipal waste, 86, 218
secondary fuels, 85, 86, 165-168, 216-218
Black lung, 41, 194

Climate change, 188, 189, 268
Coal
mining, 40-46, 191-195
ranks, 14, 15
gasification and liquefaction, 147-149
strip mining, 45, 46, 191, 192
sulfur, 198, 237, 242
Cogeneration, 106, 107
Conservation, 33, 104, 180-182, 257-260
Critical mass, 175-177

Deforestation, xx, xxiii, xxvii, 9, 79-82, 212-215, 238-241
Drilling wells, 50-55

Economic factors, 49, 236, 263, 269, 272
Efficiency
automotive, *see also* automotive fuel economy
in industry, 259-261
residential appliances, 259
Electricity
production, 179

transmission, 135
use, 180
Enhanced oil recovery, 70–73
Environmental impacts
compared to aesthetic and economic issues, 266–271
of biomass, 212–218, 239, 249, 268
of coal, 197–200, 246, 256
of geothermal, 225
of hydro, 223, 224
of nuclear, 227–233
of oil and gas, 200–210
of solar, 218–220
Externalities, 263–266

Fermentation, 165–168, 238
Fine particles, 41, 42, 77, 148, 149, 197, 211
Fischer-Tropsch (synthesis of gas from coal), 147, 148
Fission, nuclear, 175–179
Flue gas desulfurization, 199, 200
Fluidized bed combustion, 146, 147
Fuel cells, 139, 140

Geology of fossil fuel accumulations
coal, 12–17
oil and gas, 17–19
Geothermal
acquisition, 97–100
definition, 33
resource base, 35
Global warming, 188–190, 200, 256, 268
Greenhouse effect, *see* global warming

Heat
engines, 146, 158, 170, 179
space, 169, 219, 258

Hydro
conventional, 95, 223, 224, 251
wave and tidal power, 96, 225

Industrial sector conservation, 106, 107, 259-261

Kerogen (oil shale)
acquisition, 76-78
definition, 28
limitations, 77-79, 211

Land
reclamation of mined, 191, 192
subsidence, 191, 192
Less developed countries (LDCs), *see* non-industrialized countries
Liquefied natural gas (LNG), 123, 124

Mine hazards, 40-42, 193-196

Natural gas potential, 27, 35
Nuclear
acquisition, 100-104
hazards, 227-233
reactions, 175-178
resources, 33, 104

Oil (and gas)
drilling, 50-59
exploration, 46-61
production, 68-75
refining, 151-158
resources, 26-27, 35, 246-248
transportation, 114-120, 202-210

Photovoltaic cells, 125-127, 170

Pipelines, 113–118

Radioactivity, 227–230
Resources, 1–9, 35

Scale (and economies of), 272–274
Scrubber technology, 199, 200
Shale oil (*see* kerogen)
Smog, 188
Solar
  acquisition, 87–92
  active technologies, 87–90, 169, 170
  heating, 91, 92, 169
  passive technologies, 91, 92, 169
  resources, 7, 135
Spills (oil), 118–121, 202–210
Strip mining, 45, 46, 191, 192
Super conductors, 137, 251
Surface mining, *see* strip mining
Synthetic fuels, 14, 28, 77, 85, 147–149

Transmission lines (electric), 135–137
Transportation
  energy use in, 158–161, 173, 210
  of energy resources, 110–123, 135–137
Turbines, 144–146, 174, 179

Unit trains, 112

Wind
  acquisition, 93
  limitations, 171–174
  resources, 35